Real-Time Data Analytics for Large Scale Sensor Data

Advances in Ubiquitous Sensing Applications for Healthcare

Real-Time Data Analytics for Large Scale Sensor Data

Volume Six

Series Editors

Nilanjan Dey

Amira S. Ashour

Simon James Fong

Volume Editors

Himansu Das
Kalinga Institute of Industrial Technology (KIIT), Bhubaneswar, India

Nilanjan Dey
Techno India College of Technology, Kolkata, India

Valentina Emilia Balas
"Aurel Vlaicu" University of Arad, Arad, Romania

ELSEVIER

ACADEMIC PRESS
An imprint of Elsevier

Academic Press is an imprint of Elsevier
125 London Wall, London EC2Y 5AS, United Kingdom
525 B Street, Suite 1650, San Diego, CA 92101, United States
50 Hampshire Street, 5th Floor, Cambridge, MA 02139, United States
The Boulevard, Langford Lane, Kidlington, Oxford OX5 1GB, United Kingdom

Library of Congress Cataloging-in-Publication Data
A catalog record for this book is available from the Library of Congress

British Library Cataloguing-in-Publication Data
A catalogue record for this book is available from the British Library

ISBN: 978-0-12-818014-3

For information on all Academic Press publications
visit our website at https://www.elsevier.com/books-and-journals

Publisher: Mara Conner
Acquisition Editor: Chris Katsaropoulos
Editorial Project Manager: Ana Claudia A. Garcia
Production Project Manager: Punithavathy Govindaradjane
Cover Designer: Matthew Limbert

Typeset by SPi Global, India

Contents

Contributors

Christos Anagnostopoulos
School of Computing Science, University of Glasgow, Glasgow, United Kingdom

Shivaranjani Anbarsu
Department of Computer Science and Engineering, Anna University, Chennai, India

Bhargav Appasani
School of Electronics Engineering, KIIT University, Bhubaneswar, India

N.P.G. Bhavani
Department of EEE, Meenakshi College of Engineering, Chennai, India

Shanta Chakrabarty
School of Mechanical Engineering, Kalinga Institute of Industrial Technology (KIIT) Deemed to be University, Bhubaneswar, India

Everton de Matos
Pontifical Catholic University of Rio Grande do Sul (PUCRS), Porto Alegre, Brazil

Rick Siow Mong Goh
Institute of High Performance Computing, A*STAR, Fusionopolis, Singapore

Malka N. Halgamuge
Department of Elcctrical and Electronic Engineering, The University of Melbourne, Parkville, VIC, Australia

Zhaozheng He
School of Electrical & Electronic Engineering, Nanyang Technological University, Jurong West, Singapore

Fabiano Hessel
Pontifical Catholic University of Rio Grande do Sul (PUCRS), Porto Alegre, Brazil

Chittaranjan Hota
Department of Computer Science, Birla Institute of Technology and Science, Pilani, Hyderabad, India

N. Jayachitra
Department of Chemical Engineering, Dr. M.G.R. Educational & Research Institute, Chennai, India

Irosha Jayatilleka
School of Computing and Mathematics, Charles Sturt University, Melbourne, VIC, Australia

Sergio F. Johann
Pontifical Catholic University of Rio Grande do Sul (PUCRS), Porto Alegre, Brazil

V. Karthikeyan
Department of EEE, Dr. M.G.R. Educational & Research Institute, Chennai, India

Fahmida Khan
Department of Chemistry, National Institute of Technology, Raipur, India

Zengxiang Li
Institute of High Performance Computing, A*STAR, Fusionopolis, Singapore

Yongjun Li
School of Computer, Northwestern Polytechnical University, Xi'an, China

Dusmanta Kumar Mohanta
Department of Electrical and Electronics Engineering, Birla Institute of Technology, Mesra, India

Maheswata Moharana
Department of Hydro and Electrometallurgy, CSIR-Institute of Minerals and Material Technology, Bhubaneswar, India

Carlos Roberto Moratelli
Federal University of Santa Catarina (UFSC), Blumenau, Brazil

Stefanos Nikolaou
School of Computing Science, University of Glasgow, Glasgow, United Kingdom

Syam Kumar Pasupuleti
IDRBT, Hyderabad, India

Subrat Kumar Pattanayak
Department of Chemistry, National Institute of Technology, Raipur, India

Dimitrios Pezaros
School of Computing Science, University of Glasgow, Glasgow, United Kingdom

Gabriel Portal
Pontifical Catholic University of Rio Grande do Sul (PUCRS), Porto Alegre, Brazil

Purna Chandra Prusti
Cyberthings Technology, Bhubaneswar, India

Arockia Xavier Annie Rayan
Department of Computer Science and Engineering, Anna University, Chennai, India

Satya Narayan Sahu
School of Applied Sciences, Kalinga Institute of Industrial Technology (KIIT), Deemed to be University, Bhubaneswar, India

Rajesh Kumar Shrivastava
Department of Computer Science, Birla Institute of Technology and Science, Pilani, Hyderabad, India

V. Srividhya
Department of EEE, Meenakshi College of Engineering, Chennai, India

K. Sujatha
Department of EEE, Dr. M.G.R. Educational & Research Institute, Chennai, India

Ramão Tiago Tiburski
Pontifical Catholic University of Rio Grande do Sul (PUCRS), Porto Alegre, Brazil

Dheerendra Varma
University of Hyderabad, Hyderabad, India

Vettriselvi Vetrian
Department of Computer Science and Engineering, Anna University, Chennai, India

Mingzbhong Xiao
College of Information Science and Technology, Beijing Normal University, Beijing, China

Quanqing Xu
Institute of High Performance Computing, A*STAR, Fusionopolis, Singapore

Preface

Rapid developments of sensor technology and automated control technology have contributed tremendously to the building of real-time intelligence systems. With the advancements in sensor technologies and machine-to-machine communications, the paradigm of the Internet of things (IoT) has seen a major boom. Over the past few years IoT-based applications and products have become more commonplace. With the proliferation of sensor deployment, the challenge of storing and managing massive data has eventually transformed to a challenge of decision-making using the collected data. This is particularly true for real-time applications like industrial IoT, smart transportation systems, smart grids, robotics, and so on. However, with this progress a number of technological advancements are required, especially for catering to the real-time requirements imposed by most IoT-based smart environments and intelligence systems.

Most of the envisioned IoT solutions involve complex intelligence systems that have to cater to applications that are geo-distributed in nature. Examples of such IoT-based use cases include smart healthcare, management and decision-making in smart grids, and disaster management, among others. In order that the aforementioned applications can meet real-time constraints, a number of research issues need to be addressed. A potential solution is bringing processing from a central data center to much closer premises at the edge of networks through extensive distributed processing. However, this area has not been explored much. Such a computing par adigm should form the basis of large-scale deployments with real-time alarms and triggers for their control aspects. This book captures the essence of real-time IoT-based solutions that require a multi-disciplinary approach for catering to on-the-fly processing. These include methods for high-performance stream processing, adaptively streaming adjustment, uncertainty handling, and latency handling, as well as performance issues owing to geo-distributed data sources, optimization, distributed machine learning, and many others.

This book covers the following topics as they relate to real-time data analytics:

- theory and applications of hardware platforms and architectures
- development of software methods, techniques, and tools
- applications and governance
- adoption strategies for the use of massive sensor data

It also aims to provide an intellectual forum for researchers in academia, scientists, and engineers from a wide range of fields to present their latest findings and to identify future challenges in this fledging research area.

Chapter 1 investigates the state-of-the-art of the healthcare system based on IoT that mostly focuses on disease-related issues. In Chapter 2, the authors present a model for continuous, remote monitoring of the human heartbeat using a pulse sensor placed on a person's fingertip. Chapter 3 discusses methods for protecting IoT device codes from tampering and corruption, including using return-oriented programming (ROP). Chapter 4 analyzes the reliability of wireless IoT communication networks using a two-stage Monte Carlo simulation (MCS). Chapter 5 addresses the concept of Ciphertext-Policy Attribute-Based Encryption (CP-ABE) for secure data transmission and data privacy. Chapter 6 discusses combustion and presents a Long Short-Term Memory (LSTM) neural network method for predicting flame quality in an effort to reduce energy costs and hazardous emissions. Chapter 7 discusses an edge-centric, predictive methodology based on real-time model caching that decreases communication overhead and is more efficient compared to continuous raw data transmission. Chapter 8 presents the authors' proof-of-concept for a blockchain-based, decentralized application for smart building system management called Uranus. Chapter 9 highlights privacy and security mechanisms and how they can be integrated to build secure architectures for the IoT. Chapter 10 discusses solutions for using home networks with IoT devices and their impacts on software-defined networking (SDN).

Topics presented in each chapter are unique to this book and are based on the unpublished work of contributing authors. In editing this book, we attempted to bring into the discussion all the new trends and experiments in data analytics and IoT in various diversified applications. This book serves as a reference for system architects, practitioners, developers, and researchers, among others.

Himansu Das
Kalinga Institute of Industrial Technology (KIIT), Bhubaneswar, India

Nilanjan Dey
Techno India College of Technology, Kolkata, India

Valentina Emilia Balas
"Aurel Vlaicu" University of Arad, Arad, Romania

CHAPTER

Internet of Things in healthcare: Smart devices, sensors, and systems related to diseases and health conditions

1

Irosha Jayatilleka*, Malka N. Halgamuge†

**School of Computing and Mathematics, Charles Sturt University, Melbourne, VIC, Australia*

†Department of Electrical and Electronic Engineering, The University of Melbourne, Parkville, VIC, Australia

1.1 Introduction

The Internet of things (IoT) is an emerging technology in the modern information age. In the healthcare sector, a considerable number of wearables have been innovated using various types of sensors. Scientific researchers and innovators have been experimenting with wearable sensors to promote remote monitoring and early diagnosis of chronic diseases. The primary use of these sensors is to record useful parameters of the body such as temperature, heart rate (BPM), and breath acetone. Sensors are also used for storing obtained data in the cloud and local data repositories for future decision-making processes. An IoT-based healthcare system primarily contains three components: (1) cloud computing, (2) IoT, and (3) Wireless Body Area Network (WBAN).

IoT is a collection of nodes with wireless sensors that are recognized by a unique identification number. This unique number enables the communication between computers and sensors in the network to gather and distribute information [1]. Due to the increment of smart devices in a computer network, IoT devices have numerous issues, including low security, low storage volume, energy inefficiency, privacy breaches, and low scalability. Cloud computing is a way of remotely accessing shared resources in a wide network with minimal effort [2]. Most researchers are interested in cloud computing and IoT sectors since both areas are highly researched and enriched with artificial intelligence (AI) and wireless wearables with numerous sensors. While IoT empowers different devices and wearables to connect with the heterogeneous network, cloud computing allows communicating with real-world entities at any time from any place. IoT and cloud computing have the potential

Real-Time Data Analytics for Large Scale Sensor Data. https://doi.org/10.1016/B978-0-12-818014-3.00001-2

to innovate IoT healthcare devices and wearables for remote monitoring of patients' health via WBAN. The key objective of a WBAN device is measuring different parameters of the body and sending them to the cloud database. However, security and privacy are major concerns of such devices and researchers need to pay adequate attention to minimize or eliminate potential threats of IoT devices or systems before they arise [3].

Devices connected to the Internet such as implants, wearables, sensors, mHealth apps, and home monitoring tools that are rapidly increasing in number are more likely to link patients and healthcare providers in an unexpected way, utilizing big data innovatively.

The remote monitoring of health should be made trustworthy by incorporating WBAN, IoT, and cloud computing, which will be beneficial for the intelligent healthcare setting. Cloud computing also provides real-time data storage and processing for IoT devices like WBAN devices, which results in improved and cost-effective healthcare remote monitoring paradigms [4].

The key uses of an IoT healthcare system are improved clinical care, remote monitoring, and physiological and physical assistance for patients (Table 1.1).

Table 1.1 Key uses of an IoT-driven healthcare system

Clinical care	Patients who are hospitalized and require the close attention of doctors can be constantly monitored by using IoT-enabled devices. Various types of sensors can be used to collect comprehensive medical data on the status of patients in the cloud where it can be analyzed and stored before being wirelessly sent to caregivers for further analysis and review. This automated process can replace regular visits from a health professional to check patients' vital signs [5]. An IoT healthcare approach simultaneously improves the quality of care via continuous attention and minimizes the cost of care.
Remote monitoring	Lack of effective health monitoring is a key challenge faced by patients globally. IoT, cloud computing, and WBAN devices provide a sustainable solution for this challenge by enabling remote monitoring for medical professionals. An IoT healthcare system can securely capture health data via numerous sensors, apply algorithms to analyze health data, and use cloud computing for information sharing with doctors in order to make appropriate recommendations. Remote monitoring provides effective and efficient preventive measures to enhance the quality of care provided for patients [6].
Physiological and physical assistance	IoT healthcare devices largely emerged with the objective of providing assistance for patients suffering from physiological disabilities and physical infirmities. Various IoT devices and wearables have been innovated recently for these patients. As an example, people with ADHD (Attention-Deficit/Hyperactivity Disorder) benefit from IoT devices that are invented to help them develop their social skills and other activities of daily living [7].

This chapter discusses recent IoT devices and wearables for various diseases in the human body by examining the benefits, algorithms used for data analysis, identified challenges, and countermeasures for each disease. In order to do this, scientific articles on the topics of innovation of wireless wearables and IoT healthcare devices published between 2016 and 2018 were sourced from the Institute of Electrical and Electronics Engineers (IEEE) database. Most of the reviewed articles have examined different technological aspects of IoT healthcare devices. Although many types of research have applied the theory of IoT to healthcare, it is hard to find work that provides different types of devices for various types of diseases. Also, the previous work concentrates more on the technological aspects of the invented devices, which is more beneficial for professionals with knowledge and experience in the fields of IT or Engineering. This chapter targets mainly healthcare professionals such as doctors, medical researchers, and medical equipment manufacturers. IoT is still an emerging theory in the world. The complexity of the previous research makes it difficult to be understood by ordinary people. This issue is eradicated in this chapter where the complexity is quite less when compared to other similar work and helps to take the concept of IoT in healthcare to ordinary people. At the same time, the work reviewed in this chapter is the most recent inventions, making way for new inventors to develop the techniques to create more advanced or new devices that will help control more diseases.

Analysis of the devices in this chapter is categorized by the types of diseases and the affected body systems. This analysis identifies which types of diseases or body systems benefit more from the use of IoT devices. This information provides opportunities for medical professionals to collaborate with inventors to create more devices concentrating on diseases and systems that have not been targeted in previous studies.

It is essential to address security and privacy issues in IoT, as these might hold back or slow down the development of IoT in various sectors. To increase the reliability and trustworthiness of IoT, these issues should be addressed. This chapter discusses the problems and drawbacks related to the implemented devices and proposes solutions to these challenges. It also discusses the benefits of each device thus making it favorable for future use.

Most of the referred scientific publications are focused on the disease prevention and detection paradigm. The impact of the innovated systems or devices has not been discussed by most of the researchers. Another significant limitation of the research is the very few experimental events carried out for the systems or devices produced. This loophole opens a reduced transparency of the developed systems and devices to the readers, including medical researchers, scientists, medical students, medical equipment manufacturers, and all other related parties. Therefore some of the developed products under IoT healthcare might not succeed in delivering trustworthiness.

By reviewing the analysis in this chapter, future innovators in IoT healthcare systems should be able to obtain an idea of the issues that need to be addressed or minimized, which organs or medical conditions need more focus, how to enhance the benefits of the devices or systems, which parameters can be used to improve the reliability of the devices and systems, and how innovators can sustain the trustworthiness of their systems. The most beneficial parties in this research are medical practitioners, scientists, medical students, and medical device developers. The findings can be used to diagnose and control critical conditions of human diseases through the use of IoT devices in an efficient and effective manner.

The rest of this chapter is structured in the following order. Section 1.2 is a description of materials, methods, and selection criterion of acquiring the dataset for analysis. Section 1.3 presents data analysis tables and relevant graphs. We discuss our findings with relevant literature and conclude our review with some implications of our findings in Sections 1.4 and 1.5, respectively.

1.2 Material and methods

Data Collection: The raw data gathering phase was carried out by collecting 35 peer-reviewed scientific publications from 2016 to 2018. In Table 1.2, we present several health conditions and list the devices and sensors used, methods or techniques followed, benefits of using IoT devices, classifiers or algorithms used for data analysis, challenges of using IoT devices, and proposed countermeasures for given challenges. These parameters were identified as the most significant contents of the publications reviewed to be studied and analyzed.

Data Inclusion Criteria: To evaluate data inclusion criteria, we created Table 1.3, which includes a data comparison of the attributes listed previously. All publications were associated with an IoT-driven system or a device that aided a disease or a medical condition. We excluded publications that provided insufficient data for the comparison table attributes and studies that have not been published in peer-reviewed scientific publications.

Raw Data Analysis: In Table 1.4, we further categorized the reviewed articles into body systems and measured the parameters of each medical condition as well as listed benefits and challenges of using IoT healthcare devices. After the comparison was completed, three different analysis results were obtained. Later on, we summarized the analyzed data into percentages of the body systems the IoT devices are innovated for, the different measurements used in IoT devices or sensors, benefits of using IoT devices, and identified issues or challenges of the devices.

Table 1.2 provides a structured summary of the data gathered from 35 peer-reviewed publications from 2016 to 2018.

Table 1.2 IoT healthcare devices and systems for different diseases using data extraction from 33 peer-reviewed scientific publications (2016–2018)

No.	Study	Disease/ medical condition	Measured parameter	Devices/sensors used	Techniques/ methods Used	Classifiers/ algorithms	Benefits of using IoT devices	Issues/ challenges identified	Proposed solutions by authors
1	Dragan et al. (2018) [8]	Bone fractures	Injured bone electrical impedance	Accelerometer	Injured bone electrical impedance, bone temperature, pressure on fixation device, and patients' movements are measured	Naïve Bayes	Low cost	Low accuracy	Future improvements
			Bone temperature	Contact temperature sensor	Captured data classified by using three classifiers	K-nearest neighbors (KNN)	Efficiency		
			Body movements	Conductivity sensor	Data then sent to the orthopedic specialist via the Internet	Random forest	Real-time data access		
			Fixation device pressure	Force/torque sensors	After analysis of data, the doctor sends notification alerts upon the severity of the condition				
2	Rani et al. (2018) [9]	Chikungunya virus	Mosquito density	Mobile phones	Highly dense mosquito areas are identified along with other measures such as humidity, temperature, and other values	Fuzzy K-nearest neighbor (FKNN) algorithm	Efficiency	Security and privacy issues	
			Humidity temperature	Mosquito sensors	Edge servers have been located in certain areas of the city and the captured data is processed by using Chikungunya symptoms		Accuracy	Network issues	New network model has been introduced with sinks and cluster heads
					Information sent to cloud for further analysis				
3	Enshaeifar et al. (2018) [10]	Dementia	Humidity	Passive environmental sensors	A technology integrated health management (TIHM) system has been developed	Health level-7 (HL7)	Efficiency	Security and privacy issues	Data encryption methods used

Continued

Table 1.2 IoT healthcare devices and systems for different diseases using data extraction from 35 peer-reviewed scientific publications (2016–2018) *Continued*

No.	Study	Disease/ medical condition	Measured parameter	Devices/sensors used	Techniques/ methods Used	Classifiers/ algorithms	Benefits of using IoT devices	Issues/ challenges identified	Proposed solutions by authors
			Temperature	Wearable devices	Installed sensors to monitor patients' behavior at all times	FHIR4	Early diagnosis		Servers physically secured
			Appliance usage		Captured data then sent to the SME servers and caregivers can take immediate action upon the condition of a patient		Real-time data access	Data issues	Use a standard format (HL 7)
			Blood pressure						
			Pulse rate						
4	Satala et al. (2018) [11]	Sleep diseases and Freezing of Gait (FoG)	Leg movements	Smart watches	Place several wearable devices (smart bracelet, IMUs) in some places on the body of the patient		Efficiency	Security and privacy issues	
				Smart bracelets	Captured data saved in a CSV file		Real-time data access		
				Inertial Measurement Unit (IMU) Sensor Tag CC2650	Data will then be sent to a cloud for further analysis		Low cost		
				Smart Phone			Portable		
5	Afsarimanesh et al. (2018) [12]	Bone Loss	C-terminal telopeptide of type 1 collagen (CTx-1) in urine or serum.	Hioki 3536 LCR meter	Measure the level of CTx-1(C-terminal telopeptide of type 1 collagen) in human serum using a developed IoT-based biosensing device		Early diagnose	Hardware issues	
				SIGMA 6-165 centrifuge	Analyze captured data using some equations and plotting graphs to quantify the level of CTx-1		Low cost	Security and privacy issues	

				DIONEX Ultimate 3000 HPLC	Transfer the analyzed data to a cloud ThingSpeak cloud service is used		Efficiency	Network issues	
				5µ C18 100A column	Send the stored data in the cloud to the medical professionals for further analysis and investigation		Portable		
				JEOL 6480LA SEM					
				Serum CrossLaps® ELISA kit (IDS company-UK)					
				PTL-MM01 dip coater					
				Arduino Uno Wi-Fi board					
				AD5933 impedance analyzer chip					
				ADG849 switch chip					
6	Kuusik et al. (2018) [13]	Multiple sclerosis	Body movements	Developed wearable IoT sensor	Romberg balance test and lower extremity ROM-test carried out for patients	Motion signal processing-based algorithms	Ease of use	Security and privacy issues	Data has been encrypted
			Time	Android smartphone	Body movement speeds and time measured via a Bluetooth enabled sensor		Efficiency		
				Bluetooth device	Captured data via Bluetooth is analyzed using different pre-defined values (EDSS value)	European data format (EDF)	Real-time data access		
				Linux-based server	After the analysis process, the neuromuscular condition of patients sent to the smartphone		Accuracy		

Continued

Table 1.2 IoT healthcare devices and systems for different diseases using data extraction from 35 peer-reviewed scientific publications (2016–2018) *Continued*

No.	Study	Disease/ medical condition	Measured parameter	Devices/sensors used	Techniques/ methods Used	Classifiers/ algorithms	Benefits of using IoT devices	Issues/ challenges identified	Proposed solutions by authors
7	Sundaravadivel et al. (2018) [14]	Malnutrition in infants	Weight of food	Piezo sensor	Weight and time obtained via sensing board	Naïve Bayes	Efficiency	Data issues	Database approach used to increase the accuracy
			Food nutrition values from a database	Smartphone	Calculated nutrient value of the meal		Ease of use		
				Proteus System	Calculated weight of leftover food		Low cost		
			Image of the nutritional facts		Computed nutrient value of leftover food		Accuracy	Repetitive data entry	The user can customize the entry via a user interface
			Time		Provided future predictions on data				
8	Balakrishnan and Patapati (2017) [15]	Traumatic brain injury	Pressure points of the brain	Headgear attachment device with Intel Edison module	Wearable helmet used to measure angular speed, direction, and speed of the impact occurred and quantify the resulting damage		Efficiency	Data issues	A used grid of FSRs to increase the sensitivity of data
			Object velocity	Micro SD memory card	If symptoms of traumatic brain trauma are sensed an automated message will be sent to local paramedics		Real-time data access	Data issues	Machine-learning algorithms
			The angle of the hit	Portable USB power stick					
				Gyroscope					
				Tri-axial accelerometer					
				Five force sensitive resistors (FSR)					
				Smartphone					
9	Kim et al. (2017) [16]	Parkinson's disease	Foot pressure	Foot pressure sensors	Foot pressure of patients and healthy persons measured by using foot pressure sensors	Heuristic anonymization algorithm	Efficiency	Security and privacy issues	Privacy-preserving data anonymization algorithm used

				Mobile application	The comparison was made of the gravitational center according to the motions with the time when the first pressure was applied to the four sensors The pressure dissemination of the sole was compared with the strength of the pressure given to each of the four sensors The result showed the difference between the change in foot pressure and the dissemination of foot pressure among Parkinson's patients and in general Collected data sent to public data server via a developed mobile application		Accuracy		
10	Rahman et al. (2017) [17]	Diabetes	Exhaled breath acetone concentration Humidity temperature	Arduino Mega 2560 board ESP8266 Wi-Fi module FIGARO TGS 822 gas sensor DHT11 humidity and temperature sensor	Breath acetone measured with a gas sensor Humidity and temperature also measured with the gas acetone Readings sent to the Arduino via the Wi-Fi module Data stored in the SQL database Identified patients according to the ketone level Data can be accessed online via a developed web interface		Early diagnosis Real-time data access Portable Efficiency	Low accuracy	The better algorithm in calibrating the sensors

Continued

Table 1.2 IoT healthcare devices and systems for different diseases using data extraction from 35 peer-reviewed scientific publications (2016–2018) *Continued*

No.	Study	Disease/ medical condition	Measured parameter	Devices/sensors used	Techniques/ methods Used	Classifiers/ algorithms	Benefits of using IoT devices	Issues/ challenges identified	Proposed solutions by authors
11	Prouski et al. (2017) [18]	Eye stroke	Eye blood flow	Eyeglasses	Use eyeglasses with sensors to discover eye blood flow and eye strain		Efficiency		
			Eye strain	Smartphone	Send captured data to the smartphone		Portable		
				Intraocular pressure sensor (IoP)	Analyzed data with algorithms to identify patients		Real-time data access		
					If a patient in a critical condition is identified, it will immediately inform the doctor and relatives of the patient automatically		Early diagnosis		
					The color of the eyeglass changes according to the severity level of the patient and smartphone starts beeping and alerting		Low cost		
12	Yasin et al. (2017) [19]	Ventricular arrhythmia	Electrocardiography (ECG) signals	Verilog simulations with ModelSim and Synopsys tools	Measure ECG of a person with a wearable SoC (The SoC is designed with a developed VA processor with high-performance characteristics)	Quad Level Vector-based algorithm	Energy efficient	Hardware security (VA processor)	Logic locking security countermeasure has been introduced (locking a design with a secret key)
				VA processor	Wearable SoC processed and analyzed ECG features with different algorithms	Naive Bayes algorithm	Accuracy	Security and privacy issues	Advanced hardware implementation with a secret key
				Server	Extracted data transformed		Real-time data access		
					Data stored in an online server		Early diagnosis		
					Abnormal ECG features analyzed by a physician				

13	Kirtana and Lokeswari (2017) [20]	Hypertension	Heart pulses	Zigbee-based pulse sensor	Used IOIO microcontroller board to capture ECG signals of the patient		Accuracy	Security and privacy issues	
				Arduino Uno	Captured data sent to a mobile device via Bluetooth		Low cost		
				ATmega328 microcontroller	Data then sent to the FTP server called MQTT server via Filezilla		Ease of use		
				PDA	In a case of emergency SMS alert sent to the caretaker				
				FTP, application and database servers	A medical professional can download the ECG records and analyze				
14	Liu et al. (2017) [21]	Sudden Infant Death Syndrome (SIDS)	Body movements	Video camera with night vision	A video camera attached to the Artik 10 board to record movements of the infant	Eulerian magnification	Low cost	Cannot run in a real environment	Needs further improvement
				Artik 10 board	Captured video sent to a computer for Eulerian magnification process using MATLAB	Parks-McClellan algorithm	Efficiency	The system has not been tested on infants	
				Twilio Cloud service	After amplified video clip infants' chest can be easily tracked over time			Eulerian magnification takes a long time to analyze the video	More research needs to be done
				MATLAB	Frequency domain plot of the normalized video had been implemented for identifying the abnormal respiration rate of the infant and alarm generated				
					By using Twilio cloud service a text message sent to the parent or guardian immediately				

Continued

Table 1.2 IoT healthcare devices and systems for different diseases using data extraction from 35 peer-reviewed scientific publications (2016–2018) *Continued*

No.	Study	Disease/ medical condition	Measured parameter	Devices/sensors used	Techniques/ methods Used	Classifiers/ algorithms	Benefits of using IoT devices	Issues/ challenges identified	Proposed solutions by authors
15	Yatsuda et al. (2017) [22]	Heat stroke	Humidity	Raspberry Pi board	Netatmo weather station continuously measured indoor environmental temperature, humidity, CO_2 concentration, and noise level		Early diagnose	Notification issues with one LED bulb	Used series of LEDs
			Temperature	Mobile weather station	Measured data sent to the Raspberry Pi board via Wi-Fi connection			Ineffective vibration notification	Upgraded to the bracelet-type vibrator
				LED lights	Indoor heat index defined (WBGT) upon temperature and humidity Threshold value has been set			Difficult to judge the true usefulness of the system	Need further experiments
				Speakers	If the WBGT value is more than 31, alert signs will be started and elders notified with multi-sensory output devices			Hardware issues	
				Vibrator	SMS message sent to caretakers of elders when a heat stroke situation occurs				
				Netatmo weather station					
16	Narvaez et al. (2017) [23]	Heart problems	Heartrate	Pulse sensor	The pulse sensor data captured by Arduino pro mini	The High charts JavaScript library	Efficiency	Patient's activity detection not included	Future improvements
				Arduino pro mini board	Several sensor data stored in the web database		Energy efficient	Only measures BPM value of a patient	Future development with ECG and PQRST graphic
				Sim808 module	Data analyzed in the web application with real-time graphs		The minimized data size of a record	Security and privacy issues	

				GPRS antenna	Alerts are designed to activate according to the predefined BPM values		Availability of the geographical location		
				GPS antenna	Analyzed data sent to doctors via email and SMS for decision-making phase		Portable		
				LiPo battery (3.7VA 2000mAh)			Early diagnosis		
				Web Server (PythonAnyWhere)					
17	Shi et al. (2017) [24]	Autism	Proximity	Accelerometer	Two people wearing nametag-sized sensors attached to T-shirts that capture their interaction dynamics	Datavyu software	Real-time data access	Low accuracy	Place the sensor higher up on T-shirt
			Face-to-face time	Sensors	Face-to-face time, proximity, and activity level are the key parameters measured by sensors		Portable		
			Body movements	Notebook computer	Captured data then sent to a notebook computer wirelessly				
				Ultrasound sonar device	Child-specific interaction analytics and real-time reports are generated via a developed software				
					Children with Autism Spectrum Disorders (ASD) identified from the analyzed information				

Continued

Table 1.2 IoT healthcare devices and systems for different diseases using data extraction from 35 peer-reviewed scientific publications (2016–2018) *Continued*

No.	Study	Disease/ medical condition	Measured parameter	Devices/sensors used	Techniques/ methods Used	Classifiers/ algorithms	Benefits of using IoT devices	Issues/ challenges identified	Proposed solutions by authors
18	Gill et al. (2017) [25]	Gait perturbations	The orientation of the cane	Cane	A developed hybrid sensitized cane used to track the quantity and quality of physical activity	Wheatstone bridge topology	Low cost	Low battery life	Develop a battery and power subsystem
			Acceleration of the cane	Touch sensor	Raw data sent to an SD card in the cane and software via Bluetooth	multi-sensor fusion algorithm	Real-time data access	Nonlinearity of force-sensing resistors (FSR)	Replace FSR with strain gauges
			Applied load	LEDs	Both data in software and SD card uploaded to a cloud database		Ease of use	Hardware issues	
				LSM9DS1 Inertial Measurement Unit (IMU)	Raw data analyzed in the cloud by using a developed gait segmentation algorithm		Early diagnosis	Security and privacy issues	
				SGT-1A/1000-TY13 strain gauges	Caregivers and patients' relatives can remotely monitor changes in gait				
				ARM Cortex-M4F microcontroller					
				PIC16F1783 microcontroller, a DC-DC voltage regulator (by29707), battery charger (bq24041), fuel gauge (bq27542-G1), quick disconnect circuit					
19	Mandala et al. (2017) [26]	Fever	Body temperature	ESP8266 Wi-Fi modules	Body temperature measured by developed IoT-based digital thermometer known as HI-Thermo.	Fuzzy logic algorithm	Energy efficient	Unavailability of disease type	Future improvements

				P-logger	Applied fuzzy logic algorithm to determine the data sending intervals		Portable		
				INA219 Arduino Nano	Data sent to the cloud server via a Wi-Fi access point after processed by the fuzzy logic algorithm				
				DS18B20 Sensor	Doctors can observe results for decision-making				
20	Patil and Prajapat (2017) [27]	Hearing disability	Voice signals	Raspberry Pi 3	A person input voice signal to the Raspberry Pi module	Raspbian Jessie	Efficient Communication	Network issues	
			Text	SIM900A Dual-band GSM/GPRS MODULE	An API developed by using Python program to convert voice to text		Low latency	Security and privacy issues	
				Cloud server	Converted text transferred to the deaf person's mobile phone via Wi-Fi, Bluetooth, or cloud (GSM) according to the distance		Low cost		
					Similarly, the deaf person can send a text message from the mobile application to Raspberry Pi and can use the same API to convert text to voice signal		Real-time data access		
21	Pandey (2017) [28]	Stress	Heart rate	NodeMCU	Collect heart rate of a person by using a pulse sensor	Logistic regression	Low cost	Inadequate data	Future improvements
				Pulse sensor	Collected pulse records sent to the NodeMCU	Support vector machine (SVM)	Real-time data access	Low accuracy	Future improvements
				DigitalOcean cloud server	Data sent from NodeMCU to the cloud server via Wi-Fi	VF-15 algorithm		Security and privacy issues	

Continued

Table 1.2 IoT healthcare devices and systems for different diseases using data extraction from 35 peer-reviewed scientific publications (2016–2018) *Continued*

No.	Study	Disease/ medical condition	Measured parameter	Devices/sensors used	Techniques/ methods Used	Classifiers/ algorithms	Benefits of using IoT devices	Issues/ challenges identified	Proposed solutions by authors
					Data analysis carried out on the cloud server by using two classifications.	Naive Bayes algorithm		Data issues	
					Determine the person at risk or under stress with the analyzed data.				
22	Santos et al. (2017) [29]	Down syndrome (DS)	Body movements	Arduino Pro Mini with the ATMega328	Developed a wearable glove for identifying body movements	Viola-Jones algorithm	Portable	Software issues	Discarded some captured data
				HC-05 Bluetooth module	Display letters and words for literacy activities of DS patients		Low cost		
				AA NiMH batteries	Via Bluetooth, the glove can reconfigure remotely for different activities		Real-time data access		
				Neopixel stick	The smart glove can control a multimedia projector by displaying a relevant letter according to the image displayed		Ease of use		
				Laser emitter					
				Series of LEDs					
				Gobetwino software					
				AffDex SDK					
23	Valakunde et al. (2017) [30]	Maternal mortality and stillbirths	Name	Smart ASHA Android application	Accredited Social Health Activists (ASHA) are equipped with Smart ASHA Android application installed on smartphones		Real-time data access	Security and privacy issues	Used username and password

			Age	Smart ASHA website	Performed web registration of the pregnant woman by ASHA in the particular area		Ease of use		
					Via ASHA application, displaying the current week of pregnancy, current trimester, due date, high-risk patient alert, pending ANC visit if any, the status of all the four ANC visits, and the two TT doses can be obtained		Efficiency		
					Data analysis helps to minimize the maternal mortality rate and control stillbirth		Offline usability		
24	Muhammad et al. (2017) [31]	Voice pathology	Voice signals	Laryngoscope	Capture parameters via laryngoscope and stroboscope	Extreme learning machine (ELM) classifier	Ease of use	Security and privacy issues	A robust watermark is embedded into the signals
			Body temperature	Stroboscope	Data sent to the smartphone via Bluetooth by using a developed application	Discrete wavelet transform (DWT)-based algorithm	Accuracy	Dynamic scalability	Integrating different input modalities of voice
			ECG signals		A robust watermark is added, and data sent to a cloud		Efficiency		
			Humidity		Process the data using different classifiers				
					Used the ELM classifier to identify the behavior of input voice signal				
					The decision of the signal is sent to the relevant parties				

Continued

Table 1.2 IoT healthcare devices and systems for different diseases using data extraction from 35 peer-reviewed scientific publications (2016–2018) *Continued*

No.	Study	Disease/ medical condition	Measured parameter	Devices/sensors used	Techniques/ methods Used	Classifiers/ algorithms	Benefits of using IoT devices	Issues/ challenges identified	Proposed solutions by authors
25	Bhattacharya et al. (2017) [32]	Sleep apnea	Respiration temperature	Thermistor	Measure the inhalation and exhalation temperature via Thermistor sensor.		Efficiency	Security and privacy issues	
				ESP8266 Wi-Fi module	Temperature fluctuation is used to calculate the breathing rate		Low cost		
				NodeMCU microcontroller	Send the instantaneous breathing rate via an ESP8266 Wi-Fi module to a cloud		Energy efficient		
				Arduino	The smartphone app continually received the breathing rate from the cloud and doctors can monitor constantly		Accuracy		
				Microcontroller	Abnormal breathing rate is notified by the mobile application and a buzzer		Ease of use		
				Blynk cloud					
26	Ani et al. (2017) [33]	Stroke	Systolic pressure	Arduino Mega Microcontroller	Patients' health parameters are collected via sensors	K-nearest neighbors (KNN)	Real-time data access	Security and privacy issues	
			Diastolic pressure	Blood pressure sensor	Captured data sent to a cloud via Wi-Fi module	Naïve Bayes	Efficiency		
			Pulse readings	Sugar sensor	If there is a vital change in the values an SMS and an email are sent to a doctor	Random forest	Accuracy		
			Glucose level of blood	Heartbeat sensor	Finally, critical data sent to a prediction system to notify the patient of risk of stroke				

27	Vora et al. (2017) [34]	Physical therapy system for patients with motor neurons diseases	Voice signals	Smart speaker	Physical therapist sent particular exercises to a database		Efficiency	Low reliability	Future Improvements
				Speech learning unit	Smart speaker studied the exercise database and directed the patient to perform the exercise		Ease of use		
				Cloud server	Analysis based on the dialogue between the patient and the smart speaker is processed in the cloud				
					A therapist can monitor patients' various parameters via the generated report				
28	Zhou et al. (2017) [35]	Blindness	Body movements	Handhold GPRS modules	Sensors in wearing glass and walking assistant sends data to the handhold GPRS module		Efficiency	Security and privacy issues	Unique ID number embedded to core processor board
			Distance to objects	Ultrasound sensors	Obstacles are identified, and the patient informed through a speaker		Low cost		
				Wearing glass	Captured data sent to a cloud via Wi-Fi module				
				Walking assistant	Data analysis process carried out in the cloud for future usage			Devices have not been tested	
29	Huang et al. (2017) [36]	Dietary monitoring	Muscle activity of the temporalis	Electromyography (EMG) sensors	The EMG sensor captures movement frequency of temporal muscle	User-defined algorithms	Real-time data access	Low food classification accuracy	Future improvements
				Reading glass	Transmit the signals to the smartphone via Bluetooth		Energy efficient	Data issues	

Continued

Table 1.2 IoT healthcare devices and systems for different diseases using data extraction from 35 peer-reviewed scientific publications (2016–2018) *Continued*

No.	Study	Disease/ medical condition	Measured parameter	Devices/sensors used	Techniques/ methods Used	Classifiers/ algorithms	Benefits of using IoT devices	Issues/ challenges identified	Proposed solutions by authors
				Bluno Nano SD memory card	Food intake schedule, the number of chewing cycles, and broad food category can be calculated by using algorithms		Early diagnosis		
				Smartphone Health data cloud	The reading glass can provide various dietary monitoring patterns for different people				
30	Hossain and Muhammad (2017) [37]	Emotion recognition of a patient	Speech signals	Smartphone	Captured videos and speech signals via smart devices	Support vector machine (SVM)	Accuracy	Data issues	Future improvements
			Image of face	Smartwatch	Captured data processed individually in the cloudlet		Ease of use	Inefficient audio and video processing	Parallel processing for two modalities
				Smart video camera	Calculated figures compared with eNTERFACE'05 emotion database		Low cost		
				Wall-mounted camera	Identify patient's emotions and inform relevant parties if an emergency occurred				
31	Yin et al. (2016) [38]	Ocular disease	Eye images	Retinal fundus cameras	Images captured by retinal fundus cameras or portable fundus cameras	Intelligent analysis algorithms for glaucoma, age-related macular degeneration, pathological myopia, and diabetic retinopathy	Efficiency	Security and privacy issues	VPN example—Amazon VPC, Open VPN, Open Bridge Citrix
				Mobile applications and APIs	The raw data sent to a private cloud for analysis The analysis report will be sent to the public cloud with a personalized diagnosis report		Real-time data access Early diagnosis	Network issues	Encrypt all transmissions by using SSL/TLS

					with disease identified				
					Patients with high risk can make appointments with ophthalmologists with their available time slots given by the system		Ease of use		
32	Kumaresh et al. (2016) [39]	Infant mortality	Heart rate	Accelerometer sensor	Mothers' and fetal heart rates and blood pressures measured from week 5 to week 40 of pregnancy	Weibull analysis	Low cost	Data issues	Use the Weibull analysis to remove failure data
			Blood pressure	Bluetooth module	Captured data sent to a smartphone via Bluetooth		Efficiency		
					Data pattern matching carried out to identify abnormal behaviors of heartbeats		Real-time data access		
					An alert is generated and notifies the doctor if an abnormal heart rate detected.		Ease of use		
33	Raji et al. (2016) [40]	Asthma	Respiratory temperature	LM35 temperature sensor	Respiratory rate is calculated via temperature calculated by sensors	Naïve Bayes	Early diagnose	Security and privacy issues	
			Environmental temperature	Arduino ATMEGA 2560	Captured data then sent to the medical center via Wi-Fi module	J48 decision tree	Low cost		
				INA122 differential amplifier	Data analyzed using three algorithms and the most accurate respiratory rate is picked	Support vector machine (SVM)	Accuracy		
				NRF24l01 wireless module	Patients' status displayed in a web browser Once the threshold is reached an alarm is generated and the message sent to the doctor				

Continued

Table 1.2 IoT healthcare devices and systems for different diseases using data extraction from 35 peer-reviewed scientific publications (2016–2018) *Continued*

No.	Study	Disease/ medical condition	Measured parameter	Devices/sensors used	Techniques/ methods Used	Classifiers/ algorithms	Benefits of using IoT devices	Issues/ challenges identified	Proposed solutions by authors
34	Zheng and Yunus (2016) [41]	Cerebral palsy in children	Body movements	Microsoft Kinect 360 sensor system	Video and audio captured via MS 360 sensor system		Real-time data access	Low efficacy	Future improvements
				Brekel ProBody application	Captured data stored in a local database		Offline usability	Data and results are limited to the Malaysian context	Obtain data from other countries in the future
				Smartphone and PC	Raw data transmit to a cloud for data analysis process		Low cost		
					Analyzed data can be accessed by various smart devices for monitoring movements of children		Ease of use		
35	Dumripatanachod and Piyawattanametha (2016) [42]	Cervical cancer	Images of cervix cells	High-resolution microendoscope (HRME)	Images of cervix cells captured via HRME	ANN algorithms	Efficiency	HRME not used in the laboratory	Future improvements
					Sent data to Raspberry Pi 2		Accuracy		
				Raspberry Pi 2	Data transferred to a cloud-based ANN server for analysis		Low cost		
				Artificial neural network (ANN) server	Results stored in a cloud database and a copy sent to the local repository		Early diagnosis		
					Doctors can identify cancer patients				

1.3 Results

The raw data collected from 35 peer-reviewed scientific publications (2016–2018) (Table 1.2) were analyzed with respect to the body systems, measured parameters, and benefits and challenges of use. Figs. 1.1–1.4 present the graphical representations of the results.

Table 1.3 demonstrates the diseases or medical conditions related to the human body systems.

1.3.1 Human body system related to diseases

Table 1.3 Human body system related to diseases

No.	Disease/medical condition	Body system
1	Bone fractures (Dragan et al. 2018) [8]	Musculoskeletal system
2	Chickungunya virus (Rani et al. 2018) [9]	Immune system
3	Dementia (Enshaeifar et al. 2018) [10]	Central nervous system
4	Sleep diseases and freezing of gait (FoG) (Satala et al., 2018) [11]	Central nervous system
5	Bone loss (Afsarimanesh et al. 2018) [12]	Musculoskeletal system
6	Multiple sclerosis (Kuusik et al. 2018) [13]	Central nervous system
7	Malnutrition (Sundaravadivel et al. 2018) [14]	Digestive system/immune system
8	Traumatic brain injury (Balakrishnan and Patapati 2017) [15]	Central nervous system
9	Parkinson's disease (Kim et al. 2017) [16]	Central nervous system
10	Diabetes (Rahman et al. 2017) [17]	Endocrine system
11	Eye stroke (Prouski et al. 2017) [18]	Central nervous system
12	Ventricular arrhythmia (Yasin et al. 2017) [19]	Cardiovascular system
13	Hypertension (Kirtana and Lokeswari 2017) [20]	Cardiovascular system
14	Sudden infant death syndrome (SIDS) (Liu et al. 2017) [21]	Central nervous system/ respiratory system
15	Heat stroke (Yatsuda et al. 2017) [22]	Central nervous system
16	Heart problems (Narvaez et al. 2017) [23]	Cardiovascular system
17	Autism (Shi et al. 2017) [24]	Central nervous system
18	Gait perturbations (Gill et al. 2017) [25]	Central nervous system/ musculoskeletal system
19	Fever (Mandala et al. 2017) [26]	Immune system
20	Hearing disability (Patil and Prajapat 2017) [27]	Vestibular system/central nervous system

Continued

Table 1.3 Human body system related to diseases *Continued*

No.	Disease/medical condition	Body system
21	Stress (Pandey 2017) [28]	Central nervous system
22	Down syndrome (DS) (Santos et al. 2017) [29]	Central nervous system
23	Voice pathology (Muhammad et al. 2017) [31]	Central nervous system
24	Sleep apnea (Bhattacharya et al. 2017) [32]	Central nervous system
25	Stroke (Ani et al. 2017) [33]	Cardiovascular system
26	A physical therapy system for patients with motor neuron diseases (Vora et al. 2017) [34]	Central nervous system
27	Blindness (Zhou et al. 2017) [35]	Central nervous system
28	Dietary monitoring (Huang et al. 2017) [36]	Digestive system
29	Emotion recognition of a patient (Hossain and Muhammad 2017) [37]	Central nervous system
30	Ocular disease (Yin et al. 2016) [38]	Central nervous system
31	Asthma (Raji et al. 2016) [40]	Respiratory system
32	Cerebral palsy in children (Zheng and Yunus 2016) [41]	Central nervous system
33	Cervical cancer (Dumripatanachod and Piyawattanametha 2016) [42]	Reproductive system

1.3.1.1 Overview of human body systems related to diseases

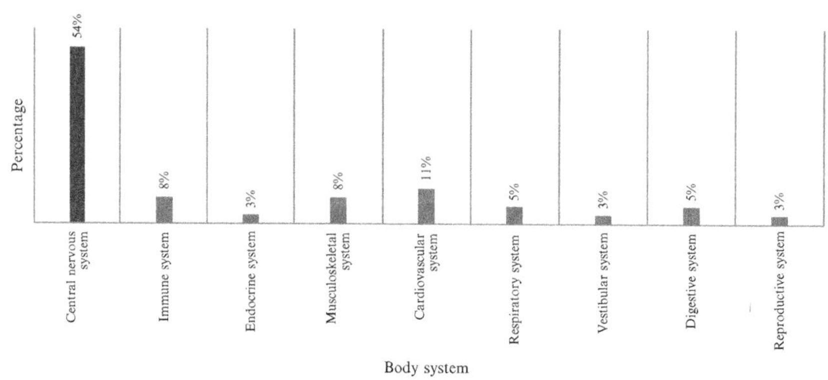

FIG. 1.1

Overview of human body systems of IoT devices for diseases using the data from 35 peer-reviewed scientific publications between 2016 and 2018 of IoT healthcare devices for diseases and medical conditions.

1.3.2 Measured body parameters, environmental parameters, and other parameters for each disease with respect to the IoT device or system

Table 1.4 Measured body parameters, environmental parameters, and other parameters for each disease with respect to the IoT device or system

No.	Disease/medical condition	Measured parameters		
		Body	**Environmental**	**Other**
1	Bone fractures (Dragan et al. 2018) [8]	Bone temperature		Injured bone electrical impedance
		Body movements		Fixation device pressure
2	Chickungunya virus (Rani et al. 2018) [9]		Humidity	
			Temperature	Mosquito density
3	Dementia (Enshaeifar et al. 2018) [10]	Blood pressure	Humidity	
		Pulse rate	Temperature	Appliance usage
4	Sleep diseases and freezing of gait (FoG) (Satala et al. 2018) [11]	Leg movements		
5	Bone loss (Afsarimanesh et al. 2018) [12]	C-terminal telopeptide of type 1 collagen (CTx-1) in urine or serum		
6	Multiple sclerosis (Kuusik et al. 2018) [13]	Body movements		Time
7	Malnutrition (Sundaravadivel et al. 2018) [14]			Weight of food
				Time
				Food nutrition values from a database
				Image of the nutritional facts
8	Traumatic brain injury (Balakrishnan and Patapati 2017) [15]	Pressure points of the brain		Object velocity
				The angle of the hit
9	Parkinson's disease (Kim et al. 2017) [16]	Foot pressure		

Continued

Table 1.4 Measured body parameters, environmental parameters, and other parameters for each disease with respect to the IoT device or system *Continued*

No.	Disease/medical condition	Measured parameters		
		Body	Environmental	Other
10	Diabetes (Rahman et al. 2017) [17]	Exhaled breath acetone concentration	Humidity Temperature	
11	Eye stroke (Prouski et al. 2017) [18]	Eye blood flow Eye strain		
12	Ventricular arrhythmia (Yasin et al. 2017) [19]	Electrocardiography (ECG) signals		
13	Hypertension (Kirtana and Lokeswari 2017) [20]	Heart pulses		
14	Sudden infant death syndrome (SIDS) (Liu et al. 2017) [21]	Body movements		
15	Heat stroke (Yatsuda et al. 2017) [22]		Humidity Temperature	
16	Heart problems (Narvaez et al. 2017) [23]	Heart rate		
17	Autism (Shi et al. 2017) [24]	Body movements		Proximity Face-to-face time
18	Gait perturbations (Gill et al. 2017) [25]			The orientation of the cane Acceleration of the cane Applied load on the cane
19	Fever (Mandala et al. 2017) [26]	Body temperature		
20	Hearing disability (Patil and Prajapat 2017) [27]	Voice signals		Text
21	Stress (Pandey 2017) [28]	Heart rate		
22	Down syndrome (DS) (Santos et al. 2017) [29]	Body movements		
23	Maternal mortality and stillbirths (Valakunde et al. 2017) [30]			Name Contact number Village ID Unique ID number

Table 1.4 Measured body parameters, environmental parameters, and other parameters for each disease with respect to the IoT device or system *Continued*

No.	Disease/medical condition	Measured parameters		
		Body	Environmental	Other
24	Voice pathology (Muhammad et al. 2017) [31]	Voice signals Body temperature ECG signals	Humidity	
25	Sleep apnea (Bhattacharya et al. 2017) [32]	Respiration temperature		
26	Stroke (Ani et al. 2017) [33]	Diastolic pressure Pulse readings Glucose level of blood Systolic pressure		
27	Physical therapy system for patients with motor neurons diseases (Vora et al. 2017) [34]	Voice signals		
28	Blindness (Zhou et al. 2017) [35]	Body movements		Distance to objects
29	Dietary monitoring (Huang et al. 2017) [36]	Muscle activity of the temporalls		
30	Emotion recognition of a patient (Hossain and Muhammad 2017) [37]	Speech signals		Image of face
31	Ocular disease (Yin et al. 2016) [38]	Eye image		
32	Infant mortality (Kumaresh et al. 2016) [39]	Heart rate Blood pressure		
33	Asthma (Raji et al. 2016) [40]	Respiratory temperature	Temperature	
34	Cerebral palsy in children (Zheng and Yunus 2016) [41]	Body movements		
35	Cervical cancer (Dumripatanachod and Piyawattanametha 2016) [42]	Images of cervix cells		

1.3.2.1 Measured body parameters, environmental parameters, and other parameters for each disease with respect to the IoT device or system

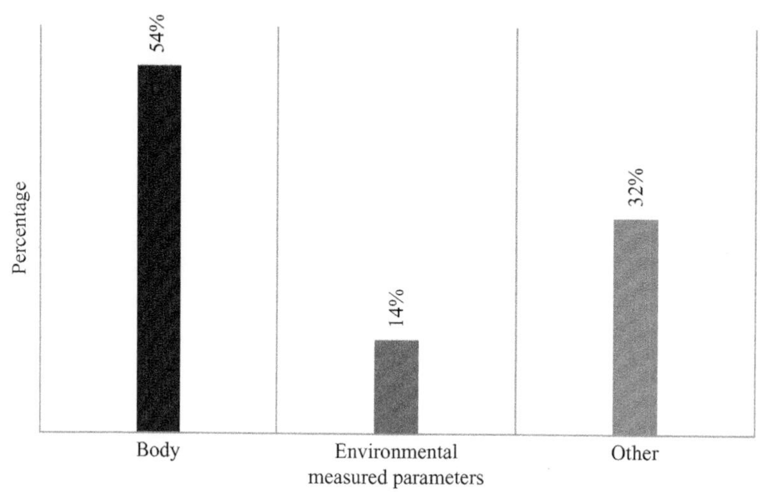

FIG. 1.2

Measured body parameters, environmental parameters, and other parameters in IoT healthcare devices using the data from 35 peer-reviewed scientific studies published in 2016–2018 of IoT healthcare devices for diseases and medical conditions.

1.3.3 Advantages of utilizing IoT-based healthcare devices or systems

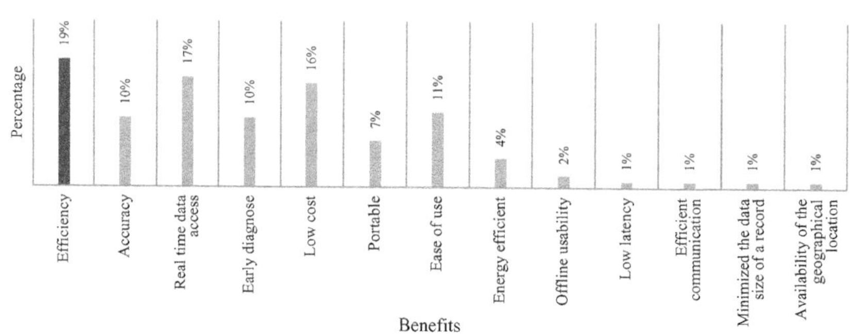

FIG. 1.3

Analysis of advantages of IoT healthcare devices using the data from 35 peer-reviewed scientific articles published in 2016–2018 of IoT healthcare devices for diseases and medical conditions.

1.3.4 Challenges of using IoT-driven healthcare systems and devices

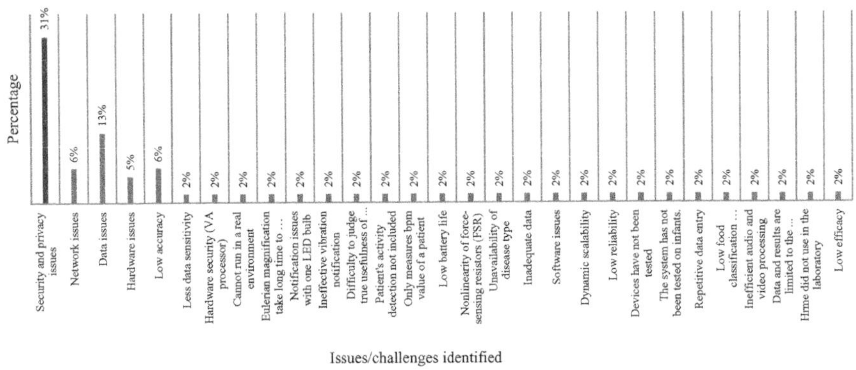

FIG. 1.4

Overview of issues and challenges of IoT healthcare devices using the data from 35 peer-reviewed scientific articles published in 2016–2018 of IoT healthcare devices for diseases and medical conditions.

1.3.5 Identified proposed countermeasures for the challenges of utilizing IoT healthcare devices and systems

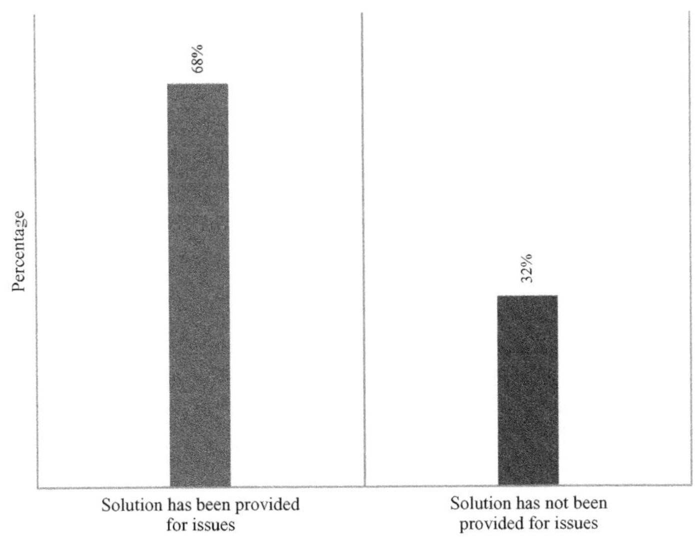

FIG. 1.5

Analysis of the proposed solutions for the issues of using of IoT devices using the data from 35 peer-reviewed scientific articles published in 2016–2018 of IoT healthcare devices for diseases and medical conditions.

Fig. 1.1 shows that 58% of the IoT devices are for diseases related to the central nervous system. These include dementia, multiple sclerosis, traumatic brain injury, Parkinson's disease, autism, eye stroke, ocular disease, and voice pathology. Approximately 13% of IoT devices are related to the cardiovascular system, including conditions such as ventricular arrhythmia, hypertension, heart problems, and stroke.

According to our results presented in Fig. 1.2, 56% of the measurements used for the IoT devices are body-related (pulse rate, body movements, voice signals, respiration temperature, foot pressure), while 16% take environmental measurements (humidity and temperature).

Figs. 1.3 and 1.4 illustrate the percentages of benefits and challenges of using the IoT devices or systems, respectively. Efficiency (32%) and real-time data access (27%) were the key advantages of most IoT devices (Fig. 1.3). The significant challenges of using IoT devices were identified as security and privacy matters (36%).

Fig. 1.5 depicts analysis on finding solutions for the challenges of using the IoT devices or systems. From the issues identified, 64% have been addressed with solutions, while 36% are yet to be resolved.

1.4 Discussion

The aim of this research is to find new IoT-driven healthcare devices, sensors, or systems to produce better information for medical professionals, thus allowing them to take steps in the prevention and detection of diseases. Furthermore our research opens a platform for data security experts to perform more research to mitigate vulnerabilities of IoT healthcare sensors and devices.

Healthcare is a key sector in the world and with the revolution of IoT, healthcare service has become more reliable and efficient. With the rapid growth of IoT devices, medical researchers and scientists have increased the detection and prevention of diseases. As a result, many IoT-driven sensors and devices have been implemented and some are already being used effectively in hospitals and for personal healthcare. These devices collect a large amount of data in a very short time period. Therefore data compromises can occur during the utilization of such devices.

Additionally, the findings of this chapter conclude that most of the IoT devices and sensors were focused on the central nervous system, and the most significant challenges of using these devices are security and privacy issues. The majority of the authors draw attention to diseases related to the central nervous system due to the ease of detectability via a device embedded with a sensor or sensors for example, [10, 15, 23, 28, 35]. The major concern of using IoT-driven devices is big data exchange, hence privacy and security challenges are relevant for many devices [9, 13, 22, 29]. Yasin et al. [19] implemented a countermeasure for hardware security by introducing logic locking security systems with a secret key embedded to the hardware component and multi-layered security infrastructures applied to ensure data security of the device.

Analysis of the reviewed studies confirmed a significant gap in most of the published articles, which is the negligence of privacy and security of end users. Only a limited number of studies [10, 13, 15, 18, 29, 30, 34, 35] have identified security measures that have been implemented to safeguard the confidentiality and integrity of the data. Moreover, there is a lack of experiments and inventions for most diseases, except for the central nervous system. Furthermore less attention has been paid to accuracy [8, 16, 23, 27, 33] and some devices have not been tested [20, 34] in real scenarios. These types of drawbacks in IoT healthcare devices and systems should not be left unanswered as the impact would be a dramatic disaster for both patient and manufacturer. An accurate and efficient estimation of sensor energy consumption is essential [43–45], for different applications running on IoT sensing devices. IoT in healthcare, especially smart devices, sensors, and systems related to diseases and health conditions are still to be explored further. Considerably more work will need to be done to resolve the issues faced in processing and storing health data with available tools, technologies, and algorithms [46, 47] related to the health system.

Future researchers should not limit their findings only to diseases of the central nervous system. They should widen their experiments to include diseases associated with the cardiovascular, immune, respiratory, musculoskeletal, and endocrine systems. Additionally, research should focus on implementing new security regulations, moderating the vulnerabilities of IoT devices through robust encryption techniques, and making people feel aware of using modern technology for smart personal healthcare.

This chapter carried out research to find new IoT-driven healthcare devices or platforms to produce better information for medical sector professionals to use. This is aimed towards taking a further step in the prevention and detection of diseases. Furthermore IoT in the healthcare sector requires continuous research since the whole world is moving forward with the concept and healthcare should not be left behind.

1.5 Conclusion

Our chapter provides a review of the latest IoT devices, sensors, and systems of smart healthcare for various diseases. In our analysis of 35 peer-reviewed publications from 2016 to 2018, we identified the types of body systems that benefit most from smart sensors or devices and the parameters measured during the data collection process. Moreover, we discussed the advantages and challenges of using such smart healthcare devices as well as proposed solutions to the challenges. Our analysis showed that a majority of the IoT devices and systems had been used to identify diseases of the central nervous system. Furthermore different body and environmental parameters are measured by these sensors for data collection. We found efficiency to be the most valuable benefit of using smart healthcare devices and systems. Simultaneously, we also found that the most significant challenge in using IoT healthcare devices is security and privacy of sensitive data. Researchers who are engaged in the field of biomedical science could base their studies and investigations on advanced IoT sensors that measure different body parameters to detect health conditions and diseases. Information security specialists and IoT-driven healthcare manufacturers are recommended to ensure the confidentiality and integrity of the information by focusing more on establishing security and privacy of the users. Future researchers can pay more attention to inventing IoT-based devices or sensors that support medical conditions related to other body systems that have not been identified in this chapter. Our research provides a clear picture to the medical, information, and communication sectors and technological innovation sector scientists of the next steps to be taken in the IoT healthcare system. Moreover, our reviewed work provides a new approach for security specialists to perform more research to mitigate the vulnerabilities of IoT healthcare devices or sensors.

References

[1] L.D. Xu, W. He, S. Li, Internet of things in industries: a survey, IEEE Trans. Ind. Inform. 10 (4) (2014) 2233–2243.

[2] Sahoo, B., Puthal, D., Mishra, S., & Swain, S. (2015). Cloud computing features, issues and challenges: a big picture. 2015 International Conference on Computational Intelligence & Networks, 116-123.

[3] W. Zhou, Y. Jia, A. Peng, Y. Zhang, P. Liu, The effect of IoT new features on security and privacy: new threats, existing solutions, and challenges yet to be solved, IEEE Internet Things J. (2018).

[4] J. Wan, C. Zou, S. Ullah, C.-F. Lai, M. Zhou, X. Wang, Cloud-Enabled Wireless Body Area Networks for Pervasive Healthcare, IEEE Netw. 27 (5) (2013) 56–61.

[5] César, H. V., & Guimarães, R. N. (2013). Systematization of clinical care based on paradigms of ubiquitous computing: a case study. 2013 8th Iberian Conference on Information Systems and Technologies (CISTI).

[6] A. Limaye, T. Adegbija, HERMIT: a benchmark suite for the Internet of medical things, IEEE Internet Things J. 5 (5) (2018) 4212–4222.

[7] Fujiwara, C. S., Aderaldo, C. M., Filho, R. H., & Chaves, D. A. (2017). The Internet of things as a helping tool in the daily life of adult patients with ADHD. GLOBECOM 2017—2017 IEEE Global Communications Conference.

[8] M. Dragan, Z. Milan, M. Milorad, V. Nikola, M. Milan, Real-time monitoring of bone fracture recovery by using aware, sensing, smart and active orthopedic devices, IEEE Internet Things J. 5 (2018) 1–8.

[9] S. Rani, S. Ahmed, S. Shah, Smart health: a novel paradigm to control the Chikungunya virus, IEEE Internet Things J. (2018).

[10] S. Enshaeifar, P. Barnaghi, S. Skillman, A. Markides, T. Elsaleh, S. Acton, et al., The Internet of Things for Dementia Care, IEEE Internet Comput. 22 (1) (2018) 8–17.

[11] Satala, P., Gaspar, V., & Butka, P. (2018). Using IoT devices for movement detection in medical environment — Proof of concept. 2018 IEEE 16th World Symposium on Applied Machine Intelligence and Informatics (SAMI), 61-66.

[12] Afsarimanesh, N., Alahi, M. E., Mukhopadhyay, S., & Kruger, M. (2018). Development of IoT-based impedometric biosensor for point-of-care monitoring of bone loss. IEEE Journal on Emerging and Selected Topics in Circuits and Systems.

[13] Kuusik, A., Alam, M., Kask, T., & Gross-Paju, K. (2018). Wearable m-assessment system for neurological disease patients. 2018 IEEE 4th World Forum on Internet of Things (WF-IoT), 201-206.

[14] Sundaravadivel, P., Kesavan, K., Kesavan, L., Mohanty, S., Kougianos, E., & Ganapathiraju, M. (2018). Smart-log: an automated, predictive nutrition monitoring system for infants through the IoT. 2018 IEEE International Conference on Consumer Electronics (ICCE).

[15] Balakrishnan, A., & Patapati, S. (2017). Automation of traumatic brain injury diagnosis through an IoT-based embedded systems framework. 2017 IEEE 8th Annual Ubiquitous Computing, Electronics and Mobile Communication Conference (UEMCON), 645-649.

[16] D. Kim, S. Hwang, M. Kim, J. Song, S. Lee, I. Kim, Development of Parkinson Patient Generated Data Collection Platform Using FHIR and IoT Devices, International Medical Informatics Association (IMIA) and IOS Press, 2017, pp. 141–145.

[17] Rahman, R., Aziz, N., Kassim, M., & Yusof, M. (2017). IoT-based personal health care monitoring device for diabetic patients. 2017 IEEE Symposium on Computer Applications & Industrial Electronics (ISCAIE), 168-173.

[18] Prouski, G., Jafari, M., & Zarrabi, H. (2017). Internet of things in eye diseases, introducing a new smart eyeglasses designed for probable dangerous pressure changes in human eyes. 2017 International Conference on Computer and Applications (ICCA), 364-368.

[19] M. Yasin, T. Tekeste, H. Saleh, B. Mohammad, O. Sinanoglu, M. Ismail, Ultra-low power, secure IoT platform for predicting cardiovascular diseases, IEEE Trans. Circuits Systems I 64 (9) (2017) 2624–2637.

[20] Kirtana, R., & Lokeswari, Y. (2017). An IoT based remote HRV monitoring system for hypertensive patients. 2017 International Conference on Computer, Communication and Signal Processing (ICCCSP).

[21] Liu, X., Takeuchi, K., Ogunfunmi, T., & Mathapathi, S. (2017). Video-based IoT baby monitor for SIDS prevention. 2017 IEEE Global Humanitarian Technology Conference (GHTC).

[22] Yatsuda, A., Haramaki, T., & Nishino, H. (2017). An unsolicited heat stroke alert system for the elderly. 2017 IEEE International Conference on Consumer Electronics - Taiwan (ICCE-TW).

[23] Narvaez, R., Villacis, D., Chalen, T., & Velasquez, W. (2017). Heart rhythm monitoring system and IoT device for people with heart problems. 2017 International Symposium on Networks, Computers and Communications (ISNCC).

[24] Shi, Y., Das, S., Douglas, S., & Biswas, S. (2017). An experimental wearable IoT for data-driven management of autism. 2017 9th International Conference on Communication Systems and Networks (COMSNETS), 468-471.

[25] Gill, S., Hearn, J., Powell, G., & Scheme, E. (2017). Design of a multi-sensor IoT-enabled assistive device for discrete and deployable gait monitoring. 2017 IEEE Healthcare Innovations and Point of Care Technologies (HI-POCT), 216-220.

[26] Mandala, S., Anggis, S., Mubarok, M., & Shamila. (2017). Energy efficient IoT thermometer based on fuzzy logic for fever monitoring. 2017 5th International Conference on Information and Communication Technology (ICoIC7).

[27] Patil, P., & Prajapat, J. (2017). Implementation of a real time communication system for deaf people using Internet of things. 2017 International Conference on Trends in Electronics and Informatics (ICEI), 313-316.

[28] Pandey, P. (2017). Machine Learning and IoT for prediction and detection of stress. 2017 17th International Conference on Computational Science and Its Applications (ICCSA).

[29] Santos, D., Szturm, D., Castro, L., Hannum, J., & Barbosa, T. (2017). Wearable device for literacy activities with people with down syndrome. 2017 IEEE MIT Undergraduate Research Technology Conference (URTC).

[30] Valakunde, N., Kumari, S., Kulkarni, P., Jumbad, S., Chavan, S., Shah, P., et al. (2017). Smart ASHA pregnancy monitoring system. 2017 International Conference on Big Data, IoT and Data Science (BID), 185-192.

[31] G. Muhammad, S. Rahman, A. Alelaiwi, A. Alamri, Smart health solution integrating IoT and cloud: a case study of voice pathology monitoring, IEEE Commun. Mag. 55 (1) (2017) 69–73.

[32] Bhattacharya, R., Bandyopadhyay, N., & Kalaivani, S. (2017). Real time Android app based respiration rate monitor. 2017 International conference of Electronics, Communication and Aerospace Technology (ICECA), 709-712.

[33] Ani, R., Krishna, S., Anju, N., Aslam, M., & Deepa, O. (2017). Iot based patient monitoring and diagnostic prediction tool using ensemble classifier. 2017 International Conference on Advances in Computing, Communications and Informatics (ICACCI), 1588-1593.

[34] Vora, J., Tanwar, S., Tyagi, S., Kumar, N., & Rodrigues, J. (2017). Home-based exercise system for patients using IoT enabled smart speaker. 2017 IEEE 19th International Conference on e-Health Networking, Applications and Services (Healthcom).

[35] Zhou, M., Li, W., & Zhou, B. (2017). An IoT system design for blind. 2017 14th Web Information Systems and Applications Conference (WISA), 90-92.

[36] Q. Huang, W. Wang, Q. Zhang, Your glasses know your diet: dietary monitoring using electromyography sensors, IEEE Internet Things J. 4 (2017) 705–712.

[37] M. Hossain, G. Muhammad, Emotion-aware connected healthcare big data towards 5G, IEEE Internet Things J. 5 (4) (2018) 2399–2406.

[38] Yin, F., Lee, B., Yow, A., Quan, Y., & Wong, D. (2016). Automatic ocular disease screening and monitoring using a hybrid cloud system. 2016 IEEE International Conference on Internet of Things (iThings) and IEEE Green Computing and Communications (GreenCom) and IEEE Cyber, Physical and Social Computing (CPSCom) and IEEE Smart Data (SmartData), 263-268.

[39] Kumaresh, S., Sabareesh, M., & Srihari, R. (2016). Non-invasive fetus heart rate and growth measurement with abnormality detection using IoT. 2016 International Conference on Electrical, Electronics, and Optimization Techniques (ICEEOT), 3655-3659.

[40] Raji, A., Devi, P. K., Jeyaseeli, P. G., & Balaganesh, N. (2016). Respiratory monitoring system for asthma patients based on IoT. 2016 Online International Conference on Green Engineering and Technologies (IC-GET).

[41] Zheng, C., & Yunus, J. (2016). The accessibility, affordability, and availability of long-term monitoring system for children with movement disorders—proposed development the Malaysian context & opportunities. 2016 IEEE EMBS Conference on Biomedical Engineering and Sciences (IECBES), 774-779.

[42] Dumripatanachod, M., & Piyawattanametha, W. (2016). Centralized pap test diagnosis with artificial neural network and internet of things. 2016 IEEE 10th International Conference on Nano/Molecular Medicine and Engineering (NANOMED), 132-135.
[43] M.N. Halgamuge, S.M. Guru, A. Jennings, Centralised strategies for cluster formation in sensor networks, in: Classification and Clustering for Knowledge Discovery, Springer-Verlag, 2005, pp. 315–334.
[44] Halgamuge, M. N., Kotagiri, R., & Zukerman, M. (2006). High powered cluster heads for extending sensor network lifetime. Proceedings of 6th IEEE International Symposium on Signal Processing and Information Technology (ISSPIT'06), Vancouver, Canada. (pp. 64-69).
[45] Halgamuge, M. N. (2007). Efficient battery management for sensor lifetime. IEEE 21st International Conference on Advanced Information Networking and Applications (AINA'07). Niagara Falls, Canada.
[46] C. Bhatt, N. Dey, A.S. Ashour, Internet of Things and Big Data Technologies for Next Generation Healthcare, Springer, 2017.
[47] C. Pradhan, H. Das, B. Naik, N. Dey, Handbook of Research on Information Security in Biomedical Signal Processing, IGI Global, Hershey, 2018.

CHAPTER

Real-time data analytics in healthcare using the Internet of Things

2

Satya Narayan Sahu*, Maheswata Moharana†, Purna Chandra Prusti‡, Shanta Chakrabarty§, Fahmida Khan¶, Subrat Kumar Pattanayak¶

**School of Applied Sciences, Kalinga Institute of Industrial Technology (KIIT), Deemed to be University, Bhubaneswar, India †Department of Hydro and Electrometallurgy, CSIR-Institute of Minerals and Material Technology, Bhubaneswar, India ‡Cyberthings Technology, Bhubaneswar, India §School of Mechanical Engineering, Kalinga Institute of Industrial Technology (KIIT) Deemed to be University, Bhubaneswar, India ¶Department of Chemistry, National Institute of Technology, Raipur, India*

2.1 Introduction

The Internet of Things (IoT) is creating a new paradigm in the arena of information technology. The growing and dynamic global network infrastructure demands increasing capabilities of the IoT, which connects a variety of physical objects and everyday items with the Internet [1]. This interconnection of physical devices and everyday objects allows the receipt and exchange of data. Radio-frequency identification (RFID) plays a crucial role in the IoT, using radio waves to read and capture information stored on a tag attached to an object.

IoT technology is swiftly gaining ground as an interesting and preferential multidisciplinary research area in various fields, including the industrial sector, healthcare systems, public transportation, and many others. The IoT in healthcare presents a fascinating technological challenge due to the increase in population and scarcity of resources [2,3]. Connecting healthcare resources such as hospitals, ambulances, doctors, nurses, assistive devices, and rehabilitation centers with patients via the IoT will enable safer and faster care at lower costs. This networking technology utilizes innovative machine-learning algorithms to assess patient health as well as to optimize the resources required by patients [4]. IoT technology in healthcare is lessening the burden on doctors and nurses to constantly collect, store, and monitor patients' health information. Such a monitoring network typically includes an intermediary processing proxy, wireless technology with integrated monitoring devices, and cloud computing to perform various

Real-Time Data Analytics for Large Scale Sensor Data. https://doi.org/10.1016/B978-0-12-818014-3.00002-4

healthcare activities such as successive sensing, monitoring, diagnosing, remote treatment, and even surgery over the Internet, by allowing for real-time data analysis, detection of critical events, and consolidation and creation of various healthcare strategies [5]. Real-time data analysis in healthcare via the IoT enables doctors and other staff to routinely assess a patient's vital signs during hospitalization, which is essential for monitoring a patient's health, minimizing mortality and morbidity, and reducing hospitalization and treatment costs. The cloud computing environment provides a huge number of computing resources such as CPU, memory, and storage to users in such a way that the users can dynamically increase or decrease their resources and their use according to their demands [6]. Researchers have studied various approaches for maximizing utilization of IoT in healthcare. For example, stored data can be used for predicting the onset of different diseases, such as diabetes mellitus [7].

Though the potential applicability of IoT in healthcare is important, some limitations and challenges need to be addressed. Mishra et al., for example, studied the challenges and demands of cloud computing [8]. Nayak et al. reported on the applicability of nature-inspired algorithms in various cloud computing problems [9]. Furthermore, mining of big data poses serious challenges [10–15]. Dey et al. [16] emphasize the analysis of real-time big data to derive actionable intelligence in enterprise applications in various domains. Previous studies reveal the connection of IoT with the next generation of healthcare [17,18]. Al-Dulaimi and Cosmas reported on how IoT affects building service managers who are tasked with maintaining health and safety systems [19]. Dhar et al. [20] proposed a prototype IoT-based method for healthcare monitoring. Pinto et al. [21] presented an IoT-ready solution for elderly living assistance that is able to monitor and register patients' vital information as well as provide mechanisms to trigger alarms in an emergency. The wristband they designed demonstrated good system performance for the implemented functionalities. Elhayatmy et al. [22] discussed details regarding wireless and sensing techniques in the application of healthcare. Dey et al. [23] discussed the IoT concept and healthcare applications to realize better healthcare with comparatively low costs. Gope and Hwang [24] addressed the major security requirements in body sensor network-based healthcare systems.

2.2 Computing system of IoT technology in healthcare activities

The IoT unites data and physical devices [25]. An algorithm based on docking studies [26–29] and molecular dynamics simulations was applied to recognizing the binding affinity of the phytochemical or compound with a target receptor, as well as for monitoring the microscopic behavior of molecules of biological interests and protein models [30–38]. Sahu et al. [39] studied the computational biology approach in management of big data in the healthcare sector. In their work, they have studied the gene interaction of Alzheimer's disease (AD) by using a string data base, binding energy, binding residue, bond name, and bond length of interaction between most targeted proteins with AD drugs through molecular docking analysis. They have tried to demonstrate some of the approaches that can be implemented for screening of genes and proteins related to human genetic disorders and collected the entire list from virtual screening, to compile a list of the most appropriate drugs for the targeted disorder. The physiological status of hospitalized patients can be recorded constantly using the IoT [40]. This can accomplish the same goals as having healthcare professionals physically observe the patient at regular intervals [41,42]. However, it is more effective in the sense that monitoring via IoT devices is continuous and the cost of care is reduced significantly.

2.3 Proposed model and its implementation

At rest, a normal heart rate is between 50 and 99 bpm [43]. During activity or excitement, the heart rate can increase to 100 times per minute or more. This occurs because the human body requires more oxygenated blood when it is active. The upper chambers of the heart—the right and left atria—receive incoming blood. The lower chambers—the right and left ventricles—pump blood out of the heart [44,45]. The circulatory system of the heart is integral for sustaining life. When the circulatory system of the heart is disturbed, circulatory disorders may occur (e.g., heart attack) [46]. Coronary artery disease can also occur as part of circulatory disorders and cause blockages in the arteries that supply blood to the heart. The electrical system of the heart coordinates the functions of the atria and ventricles. When the electrical system of the heart is disturbed, heart disease and disorders may arise. An irregular and disorganized heartbeat, either too fast or too slow, is called an arrhythmia. Millions of people experience arrhythmias [47]. Tachycardia is a heart rate >100 bpm and bradycardia is a heart rate <60 bpm. Patients suffering from arrhythmias should check their heart rates regularly.

Our remote heart-rate monitoring system is composed of four modules: (1) sensing module, (2) processing module, (3) interaction module, and (4) visualization module. Fig. 2.1 illustrates the functionality of the system. First, the sensing module collects heart-rate data from a person's fingertip. Second, the data is sent to the processing module (via an Arduino UNO microcontroller board). After the data is completely processed, it is ready to be output. The output results are sent to the two remaining modules; for current visualization the processed data is sent to the visualization module and for later or distanced visualization the data is sent to the cloud via the interaction module.

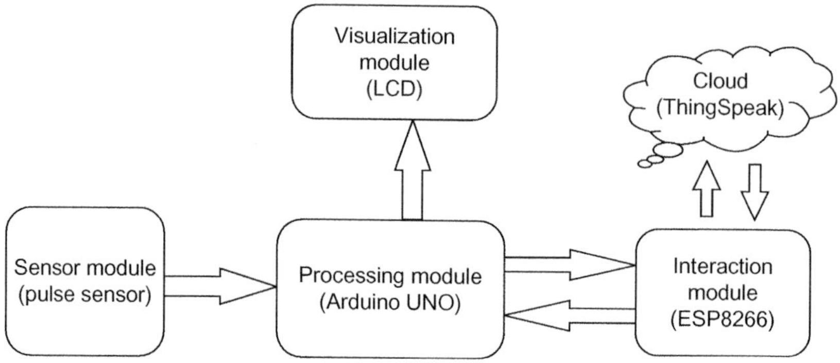

FIG. 2.1

Functional architecture of the proposed system.

2.3.1 Sensing module

The first module of our system detects heart rate via a pulse sensor attached to a person's fingertip. The sensor communicates the relative changes in intensity of light [48]. The signal value will remain constant if the amount of incident light on the sensor doesn't fluctuate. A green LED shines light into the fingertip and a phototransistor monitors or senses the light that bounces back. The sensor then produces a time-varying analog signal. The sensing module uses the photoplethysmogram (PPG) method to determine beats per minute. It provides the data coming from a particular analog pin of the Arduino board. Here analog pin A0 is connected to the sensor.

The code for the sensing module is as follows:

Sensor module code:

```
intpulsePin = 0;
signal = analog read(pulsePin);
```

Pseudocode for sensor module

```
Signal acquisition ( )
{
Set the ADC pin as A0 pin;
Read analog signal from A0 pin;
}
```

2.3.2 Processing module

As mentioned, the Arduino UNO board is used as the processing module [49,50]. It is an open source, microcontroller board based on the 8-bit ATmega328P microcontroller. The processing module reads the pulse sensor output by its analog input port. The analog signal is the process for calculating heart rate. The processing module is also connected to two more blocks: the visualization module and the interaction module. The calculated heart rate is then shared with these two modules. All the modules are controlled by the processing module.

The code for the processing module is as follows:

```
ISR(TIMER2_COMPA_vect)
{
cli();
Signal = analogRead(pulsePin);
samplecounter += 2;
int N = sample counter - lastbeat time;
```

In our study we calculated pulse rate with the help of an electronic sensor. To set the sensor we used some computational commands. The following code was used for checking the peak and trough of the pulse.

```
if(Signal < thresh && N > (IBI/5)*3)
{
If (Signal < t)
{
t = Signal;
}}
if(Signal > thresh && Signal > p)
{
p = Signal;
}
```

The following code was used for measuring heart rate in beats per minute.

```
If (N > 250)
{
if ((signal > thresh) && ( N > (IBI/5)*3))
{
pulse = true;
IBI = sample counter - lastbeat time;
Last beat time = sample counter;
If(second beat)
{
Second beat = false;
for(inti=0; i<=9; i++)
}
}
if(firstbeat)
{
First beat = false;
Second beat = true;
sei();
return;
}
word running total = 0;
for(inti=0; i<=8; i++)
```

```
{
rate [i] = rate [i+1];
running total + = rate [i];
}
rate [9] = IBI;
running total + = rate [i];
}
running total / = 10;
BPM = 60000/running total;
QS = true;
}
}
if(signal < thresh && pulse == true)
{
pulse = false;
amp = p - t;
thresh = amp/2 + t;
p = thresh;
t = thresh;
}
if (N > 2500)
{
thresh = 525;
p = 512;
t = 512;
last beat Time = sample counter;
first beat = true;
seconddbeat = false;
}
sci();
}
```

Pseudocode for processing module

```
Processing ()
{
Read data from the pulse sensor;
Keep track of the samples taken (in minutes);
Keep track of lowest point in pulse wave;
Keep track of highest point in pulse;
Calculate time period of beats in minutes (IBI);
Repeat the above steps for 10 times and calculate average IBI;
BPM = 60000/IBI;
}
```

2.3.3 Interaction module

The interaction module connects the processing unit to access the Internet. The captured data is shared to the cloud (server) using ESP8266, a low-cost Wi-Fi module with an integrated TCP/IP protocol stack. It can work as an access point and station mode. The ESP8266 module is compatible with a 3.3-V logic level. The code for the interaction module is as follows:

```
#include <SoftwareSerial.h>
SoftwareSerialesp8266(9,10);
#define SSID "NETGEAR19"
#define PASS "perfect street"
booleanconnectWiFi()
{
Serial.println("AT+CWMODE=1");
esp8266.println("AT+CWMODE=1");

delay(2000);
String cmd="AT+CWJAP=\"";
cmd+=SSID;
cmd+="\",\"";
cmd+=PASS;
cmd+="\"";
Serial.println(cmd);
esp8266.println(cmd);
delay(5000);
if(esp8266.find("OK")){
Serial.println("OK");
return true;
}
else
{
return false;
}
```

Pseudocode for interaction module (ESP8266)

```
Connect to Wi-Fi ()
{
Set ESP8266 as station mode;
Connect to a Wi-Fi access point with ssId and password;
}
```

2.3.4 Visualization module

The heart-rate data is visualized using a standard 16x2 character liquid crystal display (LCD). The data is further visualized using cloud computing. This online visualization is helpful for remote monitoring. The code for the visualization module is as follows:

Code for visualization module:

```
#include <Liquid Crystal.h>
Liquid Crystal lcd(12,11,5,4,3,2);
```

```
lcd.begin(16, 2);
lcd.setCursor(0,0);
lcd.print("Connecting...");
lcd.clear();
lcd.setCursor(0, 0);
lcd.print("BPM = ");
lcd.print(BPM);
```

Pseudocode for visualization module (LCD)

```
Display1( )
{
Initializes the LCD with 2 line and 16 Characters per line;
Set the Position the cursor at 1st row and 1st column;
Prints text (Connecting);
}
Display2()
{
Clear Display;
Set the Position the cursor at 1st row and 1st column;
Prints text (BPM);
}
```

Display1( ) code is executed on starting and Display2() code is executed after calculation of heart rate.

2.3.5 ThingSpeak

ThingSpeak is an open-source code IoT application. It can keep and retrieve data through Internet sources [51]. It provides online visualization and a text editor to perform data analysis using MATLAB. Our device communicates with the ThingSpeak application and doctors can therefore observe a patient's heart rate from any location. The code for updating with ThingSpeak is as follows:

```
#define IP "184.106.153.149"
String msg = "GET /update?key= 52Q90AFV5GHXJTJY ";
void updatebeat(){
sthingspeak.com.
String cmd = "AT+CIPSTART=\"TCP\",\"";
cmd += IP;
cmd += "\",80";
Serial.println(cmd);
esp8266.println(cmd);
delay(2000);
if(esp8266.find("Error"))
{
return;
}
```

```
cmd = msg ;
cmd += "&field1=";
cmd += BPM;
cmd += "\r\n";
Serial. print("AT+CIPSEND=");
esp8266.print ("AT+CIPSEND=");
Serial.println(cmd.length());
esp8266.println(cmd.length());
if(esp8266.find(">")){
Serial.print(cmd);
esp8266.print(cmd);
}
else{
Serial.println("AT+CIPCLOSE");
esp8266.println("AT+CIPCLOSE");
error=1;
}
}
```

Pseudocode to update with ThingSpeak

```
UpdateWithThingspeak( )
{
Create a TCP connection with thingspeak with IP address;
If error message is acknowledged return;
Update heart rate to thingspeak;
Close connection;
}
```

2.4 Working mechanism of device

The proposed system can be used in the medical field by doctors to monitor their patients' conditions from any location. The sensor device uses the PPG method, which is a way of optically obtaining a volumetric measurement of an organ, in this case the heart. The PPG uses a pulse sensor on the fingertip that measures the changes in light absorption and reflection onto the skin to measure blood flow. The blood flow is regulated by the cardiac cycle. With respect to time, each pulse appears as a peak. The generated peak and trough in the pulse count time between pulses is called the interbeat interval (IBI), which is on the order of a millisecond. For accuracy in the measurement, readings for 10 IBIs were taken and the average IBI was estimated from the collected data. Heart rate was calculated as BPM=60,000/IBI. The pulse sensor includes an LED and phototransistor, which observes the heartbeat from the fingertip. The pulse sensor output is an analog signal, so it is connected to the analog port (ADC0 pin) of the Arduino, which reads the sensor-generated data. Two serial port connections should be created. The first is used for viewing information on the terminal. For this, set the baud rate to 9600. Connect the second serial port to ESP6266 to establish an Internet connection. For this, set the baud rate to 115,200. Connect to Wi-Fi with your user ID or SSID (Wi-Fi name) and password, which is set by the router. Create a connection with the ThingSpeak server, update the heart rate, and then close the connection. The heart rate displays on the LCD that is connected to the system. Fig. 2.2 shows the complete technical flow of the device.

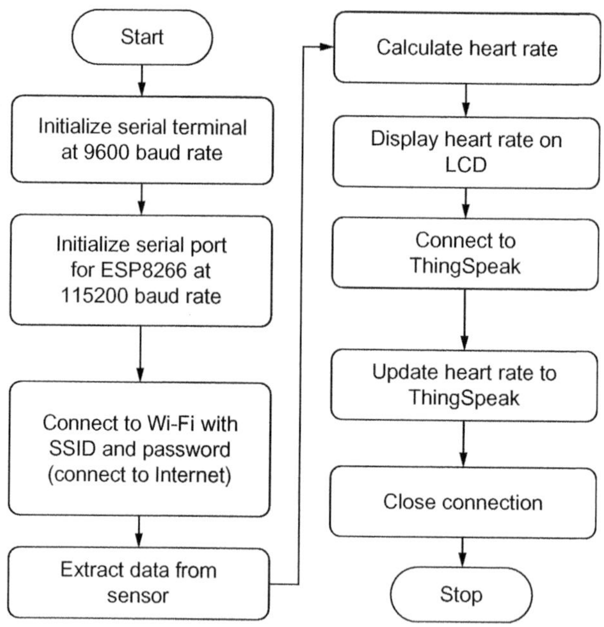

FIG. 2.2

Technical flowchart of the proposed system.

2.5 Uses and discussion of the device

Many people, regardless of health status, check their heart rate regularly. Our proposed system is one way of doing this. By using our system, a doctor can view and monitor a patient's heart rate from any location using a mobile device or laptop. Fig. 2.3 is an illustration of our device and how it works.

Although this type of device is already available on the market, ours is novel in that it uses LCD technology that can be connected to ThingSpeak via the Internet or Wi-Fi devices. The device should be registered at ThingSpeak, so that all generated heart-rate data will be automatically stored on the application's server. To view the data, one must log into the site using a registered user ID and password. This proposed system is simple and efficient and easy for both doctors and patients to use.

FIG. 2.3

Working function of device.

2.6 Conclusion

IoT technology can be used to perform various healthcare activities such as successive sensing, monitoring, diagnosing, remote treatment, and surgery over the Internet by conducting appropriate real-time data analysis. One example of this is the heart-rate monitoring system proposed in this chapter. The IoT presents new opportunities for the future of healthcare and technology.

Acknowledgments

S. Sahu and S. Chakrabarty are thankful to KIIT, Deemed to be University, Bhubaneswar. M. Moharana expresses her gratitude to the Department of Science and Technology, Government of India. S.K. Pattanayak and F. Khan are thankful to Dr. A.M. Rawani, Director, National Institute of Technology, Raipur for laboratory and library facilities.

References

[1] V.M. Rohokale, N.R. Prasad, R. Prasad, A cooperative Internet of Things (IoT) for rural healthcare monitoring and control, in: Proceedings of Wireless Vitae 2011, 2nd International Conference on Wireless Communications, Vehicular Technology, Information Theory and Aerospace & Electronic Systems Technology, IEEE, 2011, pp. 1–6.

[2] L.M.R. Tarouco, L.M. Bertholdo, L.Z. Granville, L.M.R. Arbiza, F. Carbone, M. Marotta, J.J.C. de Santanna, Internet of Things in healthcare: interoperability and security issues, in: Proceedings of the IEEE International Conference on Communications, 2012, pp. 6121–6125.

[3] P.K. Pattnaik, S.S. Rautaray, H. Das, J. Nayak, Progress in computing, analytics and networking, in: Proceedings of ICCAN 2017, Springer, 2018, p. 710.

[4] U. Kyriacos, J. Jelsma, M. James, S. Jordan, Monitoring vital signs: development of a modified early warning scoring (MEWS) system for general wards in a developing country, PLoS One 9 (1) (2014) e87073.

[5] W. Mok, W. Wang, S. Cooper, E.N.K. Ang, S.Y. Liaw, Attitudes towards vital signs monitoring in the detection of clinical deterioration: scale development and survey of ward nurses, Int. J. Qual. Health Care 27 (3) (2015) 207–213.

[6] H. Das, A.K. Jena, J.C. Badajena, C. Pradhan, R.K. Barik, Resource allocation in cooperative cloud environments, in: Progress in Computing, Analytics and Networking, Springer, Singapore, 2018, pp. 825–841.

[7] H. Das, B. Naik, H.S. Behera, Classification of diabetes mellitus disease (DMD): a data mining (DM) approach, in: Progress in Computing, Analytics and Networking, Springer, Singapore, 2018, pp. 539–549.

[8] B.S.P. Mishra, H. Das, S. Dehuri, A.K. Jagadev, Cloud Computing for Optimization: Foundations, Applications, and Challenges, Springer, 2018, p.39.

[9] J. Nayak, B. Naik, A.K. Jena, R.K. Barik, H. Das, Nature inspired optimizations in cloud computing: applications and challenges, in: Cloud Computing for Optimization: Foundations, Applications, and Challenges, Springer, Cham, 2018, pp. 1–26.

[10] A.E. Hassanien, N. Dey, S. Borra, Medical Big Data and Internet of Medical Things: Advances, Challenges and Applications, Taylor & Francis, 2019.

[11] H. Das, A.K. Jena, J. Nayak, B. Naik, H.S. Behera, H. S, A novel PSO based back propagation learning-MLP (PSO-BP-MLP) for classification, in: Computational Intelligence in Data Mining, vol. 2, Springer, New Delhi, 2015, pp. 461–471.

[12] K.H.K. Reddy, H. Das, D.S. Roy, A data aware scheme for scheduling big-data applications with SAVANNA Hadoop, in: Futures of Network, CRC Press, 2017.

[13] C. Pradhan, H. Das, B. Naik, N. Dey, Handbook of Research on Information Security in Biomedical Signal Processing, IGI Global, Hershey, PA, 20181–414, https://doi.org/10.4018/978-1-5225-5152-2.

[14] R. Sahani, C. Rout, J.C. Badajena, A.K. Jena, H. Das, Classification of intrusion detection using data mining techniques, in: Progress in Computing, Analytics and Networking, Springer, Singapore, 2018, pp. 753–764.

[15] C.R. Panigrahi, M. Tiwary, B. Pati, H. Das, Big data and cyber foraging: future scope and challenges, in: Techniques and Environments for Big Data Analysis, Springer, Cham, 2016, pp. 75–100.

[16] N. Dey, C. Bhatt, A.S. Ashour, Big Data for Remote Sensing: Visualization, Analysis and Interpretation, Springer, 2018.

[17] N. Dey, A.E. Hassanien, C. Bhatt, A.S. Ashour, S.C. Satapathy, Internet of Things and Big Data Analytics Toward Next-Generation Intelligence, Springer International Publishing, 2018.

[18] N. Dey, A.S. Ashour, C. Bhatt, Internet of things driven connected healthcare. In Internet of things and big data technologies for next generation healthcare, Springer, Cham, 2017.

[19] J. Al-Dulaimi, J. Cosmas, Smart safety & health care in cities, Proc. Comput. Sci. 98 (2016) 259–266.

[20] S.K. Dhar, S.S. Bhunia, N. Mukherjee, Interference aware scheduling of sensors in IoT enabled health-care monitoring system, in: Emerging Applications of Information Technology (EAIT), Fourth International Conference, IEEE, 2014, pp. 152–157.

[21] S. Pinto, J. Cabral, T. Gomes, We-care: an Iot-based health care system for elderly people, in: Industrial Technology (ICIT), 2017 IEEE International Conference, 2017, pp. 1378–1383.

[22] G. Elhayatmy, N. Dey, A.S. Ashour, Internet of things based wireless body area network in healthcare, in: Internet of Things and Big Data Analytics Toward Next-Generation Intelligence, Springer, Cham, 2018, pp. 3–20.

[23] N. Dey, A.S. Ashour, C. Bhatt, Internet of things driven connected healthcare, in: Internet of Things and Big Data Technologies for Next Generation Healthcare, Springer, Cham, 2017, pp. 3–12.

[24] P. Gope, T. Hwang, BNS-care: a secure IoT based modern heath care system using body sensor network, IEEE Sens. J. 16 (2016) 1368–1376.

[25] M.U. Farooq, M. Waseem, S. Mazhar, A. Khairi, T. Kamal, A review on Internet of Things (Iot), Int. J. Comput. Appl. 113 (2015) 1–7.

[26] A. Bissoyi, A.K. Singh, S.K. Pattanayak, A. Bit, S.K. Sinha, A. Patel, V. Jain, P.K. Patra, Understanding the molecular mechanism of improved proliferation and osteogenic potential of human mesenchymal stem cells grown on a polyelectrolyte complex derived from non-mulberry silk fibroin and chitosan, Biomed. Mater. 13 (2017) 015011.

[27] A. Bissoyi, S.K. Pattanayak, A. Bit, A. Patel, A.K. Singh, S.S. Behera, D. Satpathy, 2017. Alphavirus nonstructural proteases and their inhibitors. In Viral proteases and their inhibitors, Academic Press, 2017.

[28] S.M. Hiremath, A. Suvitha, N.R. Patil, C.S. Hiremath, S.S. Khemalapure, S.K. Pattanayak, V.S. Negalurmath, K.Obelannavar, Molecular structure, vibrational spectra, NMR, UV, NBO, NLO, HOMO-LUMO and molecular docking of 2-(4, 6-dimethyl-1-benzofuran-3-yl) acetic acid (2DBAA): experimental and theoretical approach, J. Mol. Struct. 1171 (2018) 362–374.

[29] S.M. Hiremath, A. Suvitha, N.R. Patil, C.S. Hiremath, S.S. Khemalapure, S.K. Pattanayak, V.S. Negalurmath, K. Obelannavar, S.J. Armakovic, S. Armakovic, Synthesis of 5-(5-methyl-benzofuran-3-ylmethyl)-3H-[1, 3, 4] oxadiazole-2-thione and investigation of its spectroscopic, reactivity, optoelectronic and drug likeness properties by combined computational and experimental approach, Spectrochim. Acta A Mol. Biomol. Spectrosc. 205 (2018) 95–110.

[30] S.K. Pattanayak, S. Chowdhuri, Effect of water on solvation structure and dynamics of ions in the peptide bond environment: importance of hydrogen bonding and dynamics of the solvents, J. Phys. Chem. B 115 (2011) 13241–13252.

[31] S.K. Pattanayak, N. Prashar, S. Chowdhuri, Effect of temperature and pressure on the structure, dynamics, and hydrogen bond properties of liquid N-methylacetamide: a molecular dynamics study, J. Chem. Phys. 134 (2011) 154506.

[32] S.K. Pattanayak, S. Chowdhuri, A molecular dynamics simulations study on the behavior of liquid N-methylacetamide in presence of NaCl: structure, dynamics and H-bond properties, J. Mol. Liquid 172 (2012) 102–109.

[33] S.K. Pattanayak, S. Chowdhuri, Size dependence of solvation structure and dynamics of ions in liquid n-methylacetamide: a molecular dynamics simulation study, J. Theor. Comput. Chem. 11 (2012) 361–377.
[34] S.K. Pattanayak, S. Chowdhuri, Effects of concentrated NaCl and KCl solutions on the behaviour of aqueous peptide bond environment: single-particle dynamics and H-bond structural relaxation, Mol. Phys. 111 (2013) 3297–3310.
[35] S.K. Pattanayak, S. Chowdhuri, Effects of methanol on the hydrogen bonding structure and dynamics in aqueous N-methylacetamide solution, J. Mol. Liquid 194 (2014) 141–148.
[36] S.K. Pattanayak, P. Chettiyankandy, S. Chowdhuri, Effects of co-solutes on the hydrogen bonding structure and dynamics in aqueous N-methylacetamide solution: a molecular dynamics simulations study, Mol. Phys. 112 (2014) 2906–2919.
[37] A. Chand, P. Chettiyankandy, S.K. Pattanayak, S. Chowdhuri, Effects of trimethylamine-N-oxide (TMAO) on aqueous N-methylacetamide solution: a comparison of different force fields of TMAO, J. Mol. Liquid 225 (2017) 926–935.
[38] S. Chowdhuri, S.K. Pattanayak, Pressure effects on the dynamics of ions and solvent molecules in liquid methanol under ambient and cold conditions: importance of solvent's H-bonding network, J. Mol. Liquid 180 (2013) 172–178.
[39] S.N. Sahu, M. Moharana, S.R. Martha, A. Bissoyi, P.K. Maharana, S.K. Pattanayak, Computational Biology Approach in Management of Big Data of Healthcare Sector, Elsevier, 2019, 247–267.
[40] A. Kulkarni, S. Sathe, Healthcare applications of the Internet of Things: a review, Int. J. Comput. Sci. Inf. Technol. 5 (2014) 6229–6232.
[41] I. Lee, O. Sokolsky, S. Chen, J. Hatcliff, E. Jee, B. Kim, A. King, M. Mullen-Fortino, S. Park, A. Roederer, K.K. Venkatasubramanian, Challenges and research directions in medical cyber–physical systems, Proc. IEEE 100 (2012) 75–90.
[42] W.B. Schwartz, Medicine and the computer: the promise and problems of change, in: Use and Impact of Computers in Clinical Medicine, Springer, New York, NY, 1970, pp. 321–335.
[43] F. Shaffer, R. McCraty, C.L. Zerr, A healthy heart is not a metronome: an integrative review of the heart's anatomy and heart rate variability, Front. Psychol. 30 (2014) 1040.
[44] B.J. Aehlert, ECGs Made Easy-E-Book, Elsevier Health Sciences, 2015.
[45] N.C. Staub, H. Nagano, M.L. Pearce, Pulmonary edema in dogs, especially the sequence of fluid accumulation in lungs, J. Appl. Physiol. 22 (1967) 227–240.
[46] A. Rozanski, J.A. Blumenthal, J. Kaplan, Impact of psychological factors on the pathogenesis of cardiovascular disease and implications for therapy, Circulation 99 (1999) 2192–2217.
[47] A.H. Lichtenstein, L.J. Appel, M. Brands, M. Carnethon, S. Daniels, H.A. Franch, B. Franklin, P. Kris-Etherton, W.S. Harris, B. Howard, N. Karanja, Diet and lifestyle recommendations revision 2006: a scientific statement from the American Heart Association, Circulation 114 (1) (2006) 82–96.
[48] S.L. Fernandes, V.P. Gurupur, N.R. Sunder, N. Arunkumar, S. Kadry, A novel nonintrusive decision support approach for heart rate measurement, Pattern Recogn. Lett. (2017), https://doi.org/10.1016/j.patrec.2017.07.002.
[49] S. Ferdoush, X. Li, Wireless sensor network system design using Raspberry Pi and Arduino for environmental monitoring applications, Proc. Comput. Sci. 34 (2014) 103–110.
[50] P.A. Pawar, Heart rate monitoring system using IR base sensor & Arduino Uno, in: Conference on 2014 IT in Business, Industry and Government (CSIBIG), IEEE, 2014.
[51] M.A. Maureira, D. Oldenhof, L. Teernstra, ThingSpeak-an API and Web Service for the Internet of Things, 2011.

CHAPTER

Lightweight code self-verification using return-oriented programming in resilient IoT

3

Rajesh Kumar Shrivastava, Chittaranjan Hota

Department of Computer Science, Birla Institute of Technology and Science, Pilani, Hyderabad, India

3.1 Introduction

The internet of things (IoT) acts as a bridge between the physical world and information technology. IoT devices are able to collect data from every nook and cranny of the physical world and process it with information technology. It is a revolutionary paradigm enabling smart applications such as traffic monitoring, smart healthcare systems, home automation, and so on [1]. The fascinating opportunities presented in this domain can only be determined through the widespread deployment of inexpensive, tiny devices in the physical world [2, 3]. The IoT makes it possible for an analyst to track environmental changes, for a grocery store to analyze shopping patterns, for a clothing manufacturer to view the changing fashion tastes and trends as they occur, and for a city administrative official to collect data from IoT devices regarding congestion and waste as well as monitor the utilization of natural resources [4, 5], among others. However, the ubiquitous deployment of IoT devices also attracts adversaries who want to intervene with malicious code [6]. Adversaries may tamper with the control flow, inject malicious payload into communication channels, and replicate devices to tamper with the overall operation of IoT systems. While a hacker controlling a video camera may evoke feelings of fear and scariness, a hacker programming the automobile brakes on a vehicle to fail or a pacemaker to stop functioning could have dire consequences [7]. It is clear that engineers, developers, and researchers must sort through an array of issues before IoT devices and their connectivity can march forward. In this chapter we envision tamper-proof code for IoT devices by using offensive techniques like return-oriented programming (ROP) [8] and defensive techniques like code checksumming.

Code verification [9] is a technique used to verify the code that is currently running on a given device with the help of a randomly instrumented code. Self-code verification is a method of code verification that investigates code integrity without

Real-Time Data Analytics for Large Scale Sensor Data. https://doi.org/10.1016/B978-0-12-818014-3.00003-6

the need of external hardware or software, such as verifying the code in memory at the time of instruction fetch without the use of any external verification program. There are numerous methods to prevent code tampering, including hashing [10], watermarking [11], label verification [12], and so on. These methods, however, do not always work in modern processors in which the code and data are processed in different caches. For example, Wurster et al. [13] showed that the method that protects code from a processor's instruction cache is easily bypassed by a code reuse attack [8]. In a code reuse attack, an adversary uploads their malicious code as data from the data cache, which is later fetched by the instruction cache and then executed.

This chapter's focus is on devising solutions to prevent code tampering in IoT devices. Our proposed method contains two layers of protection. The first layer protects code from a code reuse attack and the second layer protects the first layer as well as self-verifies the code. Recently researchers have put more emphasis on code tamper-proofing using computation-based techniques [14, 15]. Code tamper-proofing mainly includes code integrity verification by using obfuscation [10] and label verification [12] as its two major prevention techniques. These methods, however, generate high overhead in the verification process, which is not ideal for resource-constrained IoT devices with little memory. IoT devices need a code tamper-proofing method with negligible overhead. Sacham [8] showed the code prevention methods like ASLR, stack canaries, code obfuscation are vulnerable against code reuse attack. The checksumming and hashing methods can only prevent tampering attempts from the instruction cache, but ROP uses the data cache to inject malicious code into the running program. Our method prevents code tampering from the instruction cache as well as the data cache with negligible overhead. This chapter contributes a novel technique for protecting control-flow integrity (CFI) by using ROP in IoT environments. This technique works around the concept that the code that needs to be protected should also be able to verify its own integrity. Generally, in order to verify the integrity of the code more code has to be added, which leads to unnecessary overhead. We aim to minimize this overhead by using ROP along with code checksumming to develop a lightweight code self-verification technique that won't put further burden on IoT devices.

3.2 Preliminaries

3.2.1 Code tamper-proofing

Code tamper-proofing tries to prevent unauthorized modifications to the code. There are various applications that need code tamper-proofing, including the following:

- Gaming: For example, an adversary uses a cheat code to gain an undue advantage in a software game. To achieve the cheat code, an adversary attempts to tamper the code and bypass security checks [12].

- Smart IoT applications: These include applications for healthcare and traffic monitoring, for example, both of which are involved in real-time tracking [16]. An adversary can tamper with the running code to change rules or decision points, which can lead to a dangerous situation or false alarm. For instance, a patient may not receive the proper treatment in time due to incorrect information [3, 17].
- Finance: This includes software in financial sectors such as banking and the stock exchange. For example, an adversary can bypass authentication points such as login ID and password and then steal money or valuable information [7].
- System control: An adversary can use modern code-tampering techniques, such as code reuse attack, buffer overflow, and code injection, to upload malicious code to take control of a victim's system. Thereafter an adversary can launch an attack or use the victim's system for his own interests [8].

  There are various methodologies that have been proposed to avoid code tampering without external help from different resources [12, 18, 19], as we are restricted to systems that can be executed inside the program itself. The following sections discuss and compare various tamper-resistant techniques.
- Cryptography-based software protection technique [20].

This method is good for preventing static analysis and static tampering of the program. However, the running code cannot be encrypted. Hence this method is not applicable in dynamic analysis when the original code becomes available in the memory.

- Watermarking technique: This technique employs a hash function that contains a watermark signature. The hash function extracts the embedded watermark at the runtime [21]. This method is better for identifying the original owner. However, software watermarking is unable to differentiate between different copies of the original application and prevent runtime tampering.

In oblivious hashing [22], the precalculated hash value is used to check integrity of the system. If the adversary is able to bypass the checkpoints where hashes are compared or values of precalculated hashes are changed then the method is worthless.

CFI [14, 23] takes care of all control points in the code section, but ROP is a hijacking method that breaks the CFI defense.

3.2.2 Control-flow integrity

CFI prevents control flow-tampering attacks and ensures that the execution of a program follows the predefined control flow graph (CFG). Several researchers have proposed various methods to thwart control hijacking attacks [12, 15, 24, 25]. CFI protects every instruction by verifying using predefined keys. These keys are implanted before execution. Each run of instructions is validated by CFI to ensure that only valid instructions are permitted. The CFI [12] enforces the execution of the program to follow a predefined execution sequence. This prevents memory corruption, buffer overflow, and code reuse techniques such as ROP from altering the control flow. CFI mechanisms work in two steps: (1) an analysis step and (2) a verification step. The analysis step constructs the CFG, which identifies the set of legal control flow points. These CFG points are then used by a verification phase to ensure the validity and legality of the CFG. The verification is done by various methods like hashing, guards, checkpoints, and others. However, Wurster et al. [13] showed that all the methods applied can be easily bypassed by adversaries. In addition, all the existing methods for CFI generate higher overheads and therefore are not applicable in runtime environments [18, 26, 27].

3.2.3 Return-oriented programming (ROP)

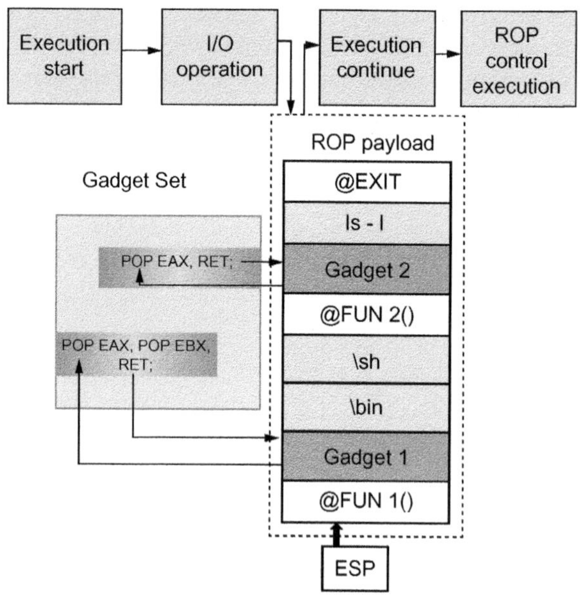

FIG. 3.1

ROP chain execution.

ROP [8, 28–30] is an adversary's weapon used to exploit a victim's code and hijack the control flow and mount a malicious code payload. ROP was originally proposed by Shacham [8] to hijack control flow with the presence of all systems security like Address Space Layout Randomization (ASLR) and Data Execution Prevention (DEP). A linked collection of gadgets is known as a ROP chain, where gadgets are short executable instruction sets terminated with a "ret" instruction. For example, the gadget "pop eax; ret;" fetches the stack value into the "eax" register and transfers control to the next instruction or gadget [31]. Stack-smashing attacks are similar to ROP. Both types of attacks use the vulnerabilities present in software, such as buffer overflow, array index out of bounds, and others. By exploiting these vulnerabilities, an adversary can mount a malicious payload into the stack. This malicious payload contains a series of gadgets with malicious system calls. Once this chain of gadgets is fetched by the instruction cache, the malicious program gets executed. An adversary may hijack a victim's system or mount an attack on another system with the help of a ROP chain [29]. For example, Fig. 3.1 shows how an adversary can hijack a terminal on a victim's system. Here, the adversary's ROP chain contains two functions. The first function supplied two arguments "\bin" and "\sh" using gadget "POP EAX, POP EBX, RET;" and the ls-1 command to second function (func2) with the help of gadget "POP EAX, RET." The func2 executes a whole string that opens a terminal and executes the ls-l command.

3.2.4 Threat model

This chapter considers the hostile host model [18] as a threat model to verify our code tamper-proofing framework. This model assumes that an adversary is able to control the execution environment and may alter the secured executable. The sole goal of an adversary is to take control of the ensured application. The challenge confronted by our code tamper-proofing strategy is to expand the endeavors required by the client to effectively prevent tampering in the secured code without considering the segments in the runtime environments.

3.3 Resilient IoT network

We used Intel Galileo Gen2 boards with the requisite Arduino-compatible sensors and Grove shields to build a small IoT testbed at our laboratory. The testbed contained IoT nodes, gateways, and a server. Sensors of the IoT nodes gathered data including light, temperature, humidity, and so on. This information is processed by gateways and a server. The server makes smart decisions based on feedback from the sensors. For example, for an IoT smart home device, if there is a sudden rise in temperature, the server will activate a building's air conditioning. This chapter presents a secure, tamper-proof, and resilient IoT framework to help IoT devices perform various services without interruption and, most importantly, carry out mechanisms that help remedy any problems or anomalies observed by the readings. We began by creating a web application uploaded on a sample Intel Galileo Gen2 board and accessed through a computer to observe the values of the various parameters brought to us by the sensors attached to the board. Extending this idea forward, we set up a network infrastructure of the boards allowing them to communicate with each other.

The network of clients (in this case, the boards), gateways, and server collate the data being retrieved from the sensors. The data passes through gateways where it is filtered before being stored in the server for further use. A similar type of testbed has been used by other researchers for different applications like healthcare via IoT [16, 17, 32]. The system consists of a sensor layer, a gateway layer, and a server

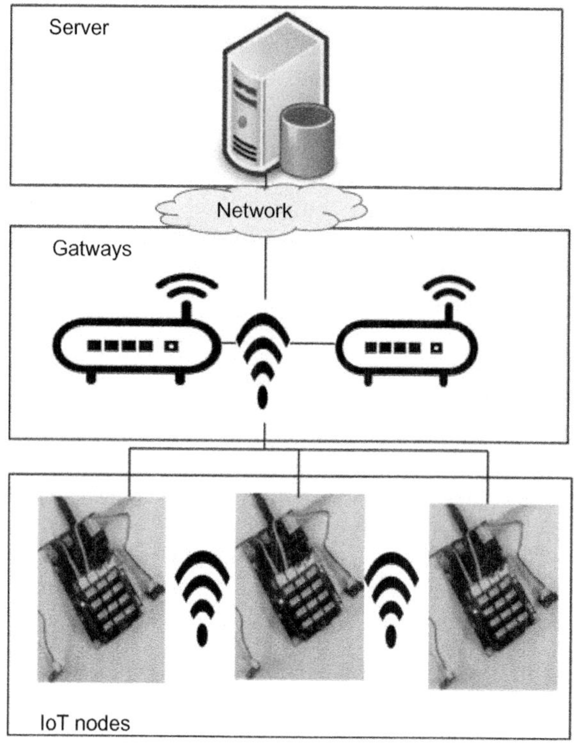

FIG. 3.2

Resilient IoT architecture.

layer. The gateway layer is designed to process the incoming readings and filter them into two categories: (1) relevant and (2) irrelevant. Only the relevant readings from the nodes end up reaching the server for display. The gateway filters the readings using a windowed averaging filter, which gives an average estimate of the values within the window. The server is responsible for carrying out the process of setting thresholds for the values as per the requirements of the user. The user is provided with an option of setting his own threshold limits for the entire system, which are then conveyed to the gateway for updating its settings accordingly. Fig. 3.2 explains the setup of the testbed and the flow of readings from nodes to the user. There are some conjunction points in the testbed that can interrupt the communication of IoT nodes. The implemented IoT network uses a DHCP server to generate a dynamic IP address for each node. The DHCP server updates IP addresses every 2–3 days. If a new node joins the network, the server automatically assigns a unique IP address to that node. This dynamic update creates a challenging situation for the IoT network, however. The IP addresses of the server and gateway are also regenerated every 2–3 days therefore interrupting the communication between the node and the gateway and server.

3.3.1 Resilient server

Algorithm 1 represents server activity and explains how the server becomes resilient in this dynamic IP change scenario. The IoT server is the main source for maintaining resiliency in the IoT network. Algorithm 1 runs in a server machine where the server maintains a list of all active gateways. At any time only one gateway becomes a master and communicates with all IoT nodes belonging to the same network. If the IP address of the gateway changes or the gateway physically gets damaged, Algorithm 1 looks for a list of dormant gateways from which it selects the next available one and makes it active. In case there is no active gateway found by the server, the server broadcasts a message to all nodes in the network. Hence, all IoT nodes send their sensor values directly to the server until the gateway is dead.

Algorithm 1 Resilient server algorithm

```
Input: G: list of gateways [g1, g2..., gn], status
while (i≤n) do
    if (status received from g[i]) then
        if (Gateway IP is not set) then
            Push the IP address of Gateway;
            Store its status into G;
        else
            Update status;
        end if
        if (status =="active") then
            Read values and make it master;
      else
            g[i] ← dormant
      end if
  end if
end while
```

3.3.2 Resilient gateway

Algorithm 2 Resilient gateway selection algorithm

```
Input: G: list of gateways [g1, g2..., gn], counter, status
while (i≤ n) do
    status message received from Server
    if (status == "active") then
        Start broadcasting IP to IoT node;
    else
        Wait ();
    end if
end while
if (counter == 0) then
    goto step 2
end if
```

Algorithm 2 shows selection of an active gateway from the available gateway list. It is the job of the gateway to send a heartbeat to the server. By receiving this heartbeat, the server adds a gateway to the active server list. Whenever the server's IP address is updated, the server broadcasts the address into the network so that all nodes in the network update to the same (the server's) IP. Algorithm 2 depicts this process. Once a gateway is assigned as active it broadcasts a message to the network so all IoT nodes update their IP addresses accordingly, ensuring communication is maintained.

3.3.3 Resilient IoT node

Resiliency in IoT nodes needs a method by which an IoT node is able to run its services without interruption. For this purpose we propose a lightweight code tamper-proofing framework that allows an IoT node to execute its services even in hostile host environments.

3.4 Code tamper-proofing

Since IoT devices are resource-constrained, they require protection methods with low overhead. Here we use a ROP chain-based code-verification method to thwart code hijacking attacks [18, 26]. A ROP chain is a systematic arrangement of gadgets and selected control points. The proposed method uses a ROP chain to prevent code tampering. In this method we interleave addresses of functions with gadgets and upload them into the stack. The main program calls this ROP chain by using a verification function and executes the code from the data section. The control flow sequence is absolutely hidden from an adversary and is protected by using the ROP chain. If an adversary tries to replace the ROP chain then the program will not execute. To frame our ROP chain, we used a set of randomly selected gadgets, which increases the ROP chain's strength. This code tamper-proofing method uses another layer in addition to the ROP chain. The second layer is the code guard layer, which prevents code tampering from the code execution section. This second layer uses dynamic hash of instructions. If an adversary tries to tamper with the code segment the guard will invalidate the hash value as well as change the value of memory addresses used by the ROP chain and hence the program will halt. Our method for building a ROP chain is comprised of the following steps.

3.4.1 Gadget identification

Algorithm 3. Algorithm to select minimal gadget set

Input: S: Gadget database
Output: Set of gadgets with execution time.
1: gadget = fetch first gadget from database;
2: while (gadget) do

```
    a = 0;
    b = 0;
    Time[a] = 0;
    Generate ROP chain with selected gadget;
    while b < 100 do
        Time[a]+ = time(ROP chain); // Execution time each gadget.
        b = b+1;
    end while
    avg = Time[a]/100; // Taking average of execution time.
    a=a+1;
    gadget = fetch next gadget from database;
end while
return List of all gadgets with execution time;
```

Gadgets refer to a set of short instructions ending with a return (ret) instruction. To search gadgets in a binary, first we generate the assembly file from the executable using the "objdump" command. Then we search for a "ret" statement in that file. After finding the "ret" we traverse back and store all combinations with "ret," such as "pop ebx; ret" and "pop eax; pop ebx; ret;". A gadget search algorithm collects all possible combination of gadgets into a database. This database is huge and contains some useless gadgets. Algorithm 3 is used to filter the gadget database. Algorithm 3 selects each gadget from the database and creates an executable ROP chain with one gadget. Each chain is executed 100 times to ensure the gadget is executed correctly and then stored as an executable gadget into the database. The main task of the ROP chain is to run all possible combinations of gadgets and determine the minimum execution time of each ROP chain. Finally, all gadgets are executed with a Genetic Algorithm (GA) to search for the best set of gadgets with minimal overhead.

3.4.2 Gadget set selection using GA

Algorithm 4. GA to find preferred gadgets

```
Input: S: database populated by Algorithm 3
Output: Preferred ROPchain
Time = 0;
// Populate initial population of chromosome from database S.
    Initialize population set, s [ ] ← s1, s2,...,sn;
Select initial chromosome based on execution time taken by gadget;
while time (ROPchain) ≤ time (Original code) do
    // Evaluate fitness value of ROPchain;
    timeTaken = endtime() − starttime();
    Select new chromosome as parent to reproduce based on fitness;
    Crossover parents and Produce the offspring
    use this offspring and execute ROPchain;
    Mutate individuals, based on fitness function;
    time = time + 1;
end while
return Preferred ROPchain.
```

To optimize a ROP chain that has less overhead in terms of time, we use the GA-based search algorithm [27]. Algorithm 4 helps us to select the best gadget set to create a ROP chain with minimum execution time. The input chromosomes are populated by Algorithm 3. These gadgets are selected on the basis of minimum execution time. Algorithm 4 uses GA to identify the preferred ROP chain. Algorithm 4 needs chromosomes for initial population of candidate solutions. The chromosome is a collection of genes (provided by Algorithm 3). Algorithm 4 selects two chromosomes based on least execution time of the genes (gadgets). After selecting the initial population, the next step is to evaluate the fitness of each chromosome. The fitness function calculates the time taken by each chromosome in the ROP chain. If execution time of the ROP chain is less than or equal to the threshold (actual execution time of the executable file) value, then this chromosome may generate a preferred ROP chain.

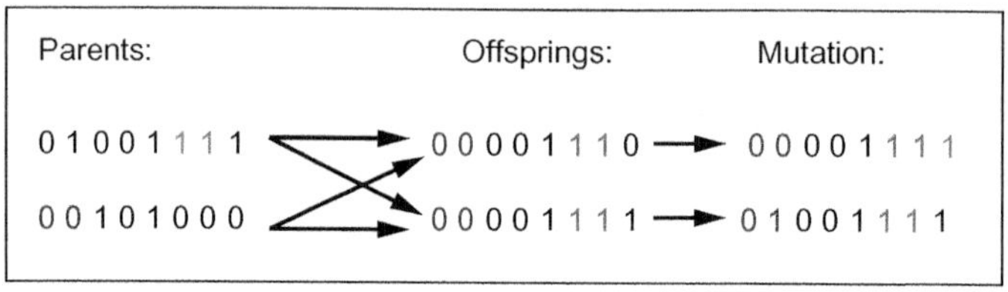

FIG. 3.3

Execution of GA.

On the basis of fitness, the selection of a new parent is carried out. The crossover of the new parent generates offspring. Fig. 3.3 represents the GA process and provides an example in which two initial populations (parents) are represented by two 8-bit chromosomes. The crossover generates the offspring with inherited genes like 11 and 00 from chromosome 1 and chromosome 2, respectively. The mutation function from Algorithm 4 randomly changes the values of some genes in the new population. Algorithm 4 continuously produces offspring until the termination condition occurs, that is, less or equal execution time of the ROP chain in comparison to the binary without the ROP chain.

3.4.3 ROP chain generation

The ROP chain intermixes gadgets with the binary code of a program to harden code tampering. To generate a ROP chain we need "N" number of gadgets and "M" number of functions that we want to protect. Now our method arranges functions and gadgets in a sequence, so that after execution of each gadget it returns the control sequence to a function as per the control flow. All the argument of function is passed to the function with the help of gadgets as shown in Fig. 3.4. Since altering the ROP chain makes the translated instructions to segmentation fault, this self-verifiably checks the integrity of the secured code. Thus we use the term self-verification with the ROP chain. For example, Fig. 3.4 shows the process to convert the original program into a ROP chain. Here, the function f1() is now converted into a ROP chain. Listing 3.1 represents a sample ROP chain. This ROP chain contains the address of function f1() with the random numbers of ROP gadgets. This translated ROP chain is responsible for executing function f1() and verifying it at the time of execution. If an adversary tries to change a single byte in this ROP chain, the entire chain malfunctions and execution halts.

```
void verify() {
    char *ROP="XXXXXXXXXXXXXXYYYYYY\x6b\xc9
    \x05\x08\x56\xb8\x1c\x08\x2c\xbc\x2a\x08verify\x8c
    \xb1\x07\x08\x5e\xae\x23\x08\x3b\xc6\xb6\x08\x72
    \xca\x24\x08\x0b\xae\x36\x08\x04\xbe\x0c\x08
    \xdf\xac\xbe\xdc\x50\xc7\x06\x08";
    char buf[6];
    strcpy(buf,ROP);
}
```

LISTING 3.1

ROP chain.

Fig. 3.4 shows the conversion of f1() into a ROP chain by using various gadgets suggested by Algorithm 4. ROP(f1) is the main function of the program. Once the control reaches the verify function ROP(f1), the ROP chain that we've constructed is copied into the buffer. Since the ROP chain size exceeds the buffer size, adjacent memory locations are overwritten and we are able to manipulate the instruction pointer according to our liking. The ROP chain starts with the address of the first gadget, so it executes and transfers control to next gadget. In this chain one of the gadgets is used to execute f1() and saves the return address in the stack.

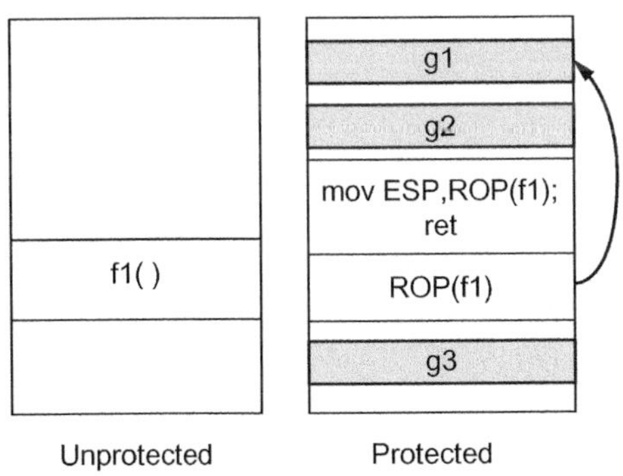

FIG. 3.4

ROP protection.

Gadgets are executed one after another, transferring the control sequence to the next following the predefined control flow. If a single digit in the ROP chain is tampered with then the address of the next instruction is changed, and the availability of the next gadget is affected; hence, execution terminates. Thus we refer to this ROP chain as the verification code. To prevent this ROP chain we deploy another layer of security (i.e., check-summing), which also protects the code against tampering attempts.

3.4.4 Code checksumming

During compilation, the code written in a high-level language (HLL) gets converted into an assembly language from which it is finally converted into a binary that can be easily interpreted by the processor. Every code line in the assembly language is interpreted as a string of opcodes, which in turn lets the processor know exactly what needs to be done. This is the underlying principle that code checksumming works on. With this technique each code line to be protected is interpreted as a string of opcodes and added up, which in the end gives a checksum value in hexadecimal format. It verifies that the checksum value calculated is the same as that of the untampered code section as well as verifies the integrity of the code that is protected. Since this is a tamper-resistant technique it should be able to verify the integrity of the code that it is protecting (code signature) as well as take appropriate actions when tampering is discovered. The whole mechanism can be divided into a detection part and an action part.

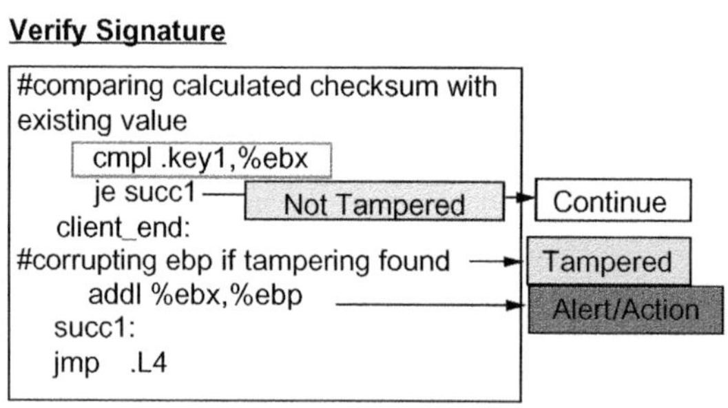

FIG. 3.5

Protection by guards.

The detection part should be able to recognize modifications in the code and trigger the action part [9]. The detection part can be further divided into two steps. Firstly, we need to obtain a signature of the critical code that needs to be protected. Secondly, we have to verify the integrity of the signature to ensure that the critical piece of code hasn't been tampered with in any way. In our case, the detection is done using the code segments known as guards, which checksum the piece of code that needs to be protected and carry out the appropriate actions if they find out that the code has been tampered with see Fig. 3.5.

3.4.5 Cross-verifying guard framework

Software cracking is a prevalent problem in today's world. It is an exploit in which a malicious entity obtains a copy of software and is successful in breaking its protection mechanism therefore allowing the entity to make modified copies of the program. In order to make software cracking inherently difficult and time-consuming, Chen et al. [10] proposed a network of cross-verifying checksum guards that raises the bar for would-be attackers and makes it harder for them to exploit the code. In this network, the guards responsible for checksumming the code protect each other as a part of strongly connected network. This network is harder to bypass because the security is distributed among all the guards. Moreover, there are many permutations and combinations that can be applied while making a guard network, which make it even harder for a malicious entity to penetrate. The level of security can always be customized by the user by simply adjusting the number of guards that will form a network. A guard is a piece of code responsible for checksumming a piece of code and performing predefined actions if tampering is observed. In our case, the appropriate action that has been predefined for the guard to take is to corrupt the stack pointer so that the program cannot progress any further. However, the severity of actions can be varied, ranging from the mildest of simply making a log of the event to the extreme of corrupting the stack pointer. If no tampering is observed then the program execution proceeds as intended. A group of guards can provide a more resilient form of protection as opposed to a single guard. Some advantages of having multiple guards are as follows:

- If a program has multiple regions of code that need to be protected then multiple checksumming guards can be deployed to protect the critical sections of the code.
- There is no single point of failure for a guard network since the guards are invoked at different points in the runtime, so it is not that easy to detach the critical code from these checksumming guards.
- There are multitude numbers of ways in which a strongly connected guard network can be configured. Different kinds of formations can be formed, making it harder for a hacker to mount an attack.
- Based on the level of protection required, the number of guards can be increased accordingly.

3.4.6 Code checksumming with ROP

Though ROP in itself is an efficient way to passively identify when someone has tampered with your program, as a standalone technique it has certain vulnerabilities. One of these is that the attacker might identify the verification code where our ROP functionality had been implemented and circumvent it completely or reverse engineer it and mimic whatever was being done by our ROP code in a HLL. If the attacker is successful in doing either of these things, then the malicious entity can go forward and tamper with the critical piece of code that the ROP chain was protecting. In order to avert this kind of situation it would be useful to deploy code checksumming in the form of a cross-verifying guard network over the critical pieces of code in conjunction with ROP functionality. There are certain advantages to implementing these two techniques in conjunction with one another. For example:

- There is no single point of failure. If the attacker is able to circumvent the ROP functionality then he still has to bypass the code checksumming guard network before he can tamper with the critical code. Similarly, just bypassing the guard network and tampering with the critical code won't be of any help because our ROP functionality will identify that the code has been tampered with and perform the appropriate actions.

- The only way for an attacker to bypass this is to simultaneously disable the guard network as well as the ROP functionality, which will be time-consuming.
- The protection provided by the cross-verifying guard network extends even outside the code region to be protected. The addresses of the flags are changed inside the protected region of code if someone tampers with it outside of the protected region. The guard network will detect this because of the checksum calculations.
- Putting these cross-verifying checksum guards into the assembly-level code instead of into the source code is useful because it makes it more difficult to decipher, as both the guards and the ROP functionality have been implemented. This adds a layer of code obfuscation to the program as well.

Let us elaborate further on the steps that need to be followed in order to implement ROP with code checksum on a program for enhanced security.

- Firstly, select a verification function from the given code and implement that functionality using ROP and the gadgets from the critical pieces of code that need to be protected.
- Secondly, generate an assembly level code for the modified source code infused with ROP.
- Thirdly, implement a cross-verifying checksum guard network protecting the critical piece of code. This will need to make appropriate modifications to the ROP chain that you've used since the addresses of the gadgets will change.
- Finally, place XOR-encrypted values of the correct checksum values (key) with predecided values on the assembly level code so as to ensure no one tampers with the keys. Take appropriate actions if tampering with the keys is evident.

```
movl   .key, %edi
xorl   .test, %edi
cmpl   %edi, . ans
je     guard
addl   %ebx, %ebl
guard:
```

LISTING 3.2

XOR encryption.

For experimental purposes we implemented a simple two-guard, strongly connected network that worked in conjunction with our ROP functionality. Fig. 3.6 shows the sample guard template we used in our program.

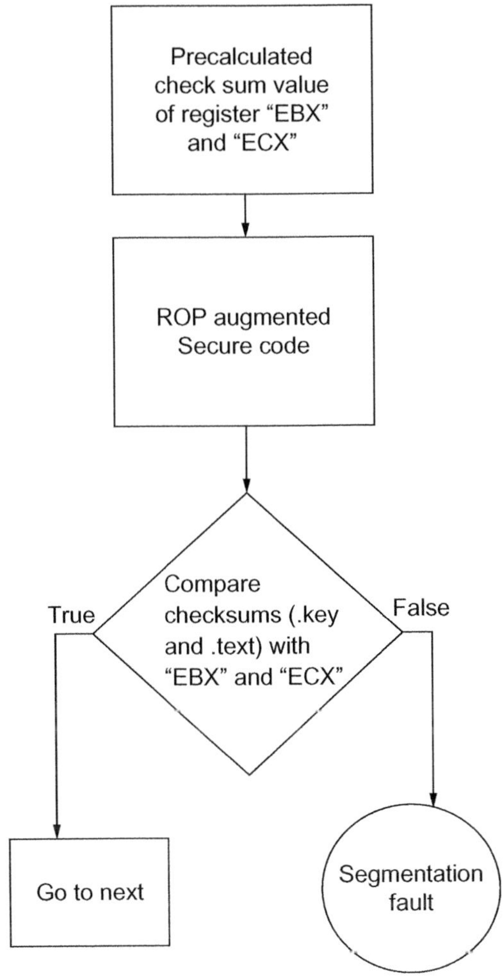

FIG. 3.6

Flow chart of guard template.

Listing 3.2 shows how the XOR-encrypted checksum tester works. In order to ensure no one tampers with the keys that have been put in the assembly-level code, XOR testers, whose template is given in Listing 3.2, can be inserted. Here, ".key" is the precalculated key for a given piece of code, ".test" is an arbitrary value with which the XOR of ".key" is calculated, and ".ans" is the correct XOR encrypted value. If ".key" XOR ".test" has a different outcome than ".ans" it means the precalculated key has been tampered with. Finally, create the executable, and now you have a tamper-proof executable protected by ROP and code checksumming.

3.5 Experimental result

3.5.1 Tamper resistance

To circumvent these two layers of security, an adversary needs to disable both layers at the same time. However, this is infeasible because the first layer ROP chain uses gadgets that are already available in the binary.

Each gadget is associated with an address and tampering with the gadgets leads to segmentation fault. The second layer calculates dynamic hash using registers value, so any tampering in instructions register value alerts the guard template. Layer one protects the data segment from tampering and layer two protects the code segment of the binary.

3.5.2 Performance overhead

Table 3.1 lists the performance overhead published by other researchers. We test performance overhead using an IoT testbed and Ubuntu 16.04 desktop system. Table 3.2 shows the results of our Algorithm 4, which generates a ROP chain with various gadget sets (i.e., chromosome is executed 100 times). Table 3.2 shows the number of times a gadget appears in the ROP chain with its average execution time. Fig. 3.7 shows a comparative study between the ROP chain suggested by Algorithm 4 and a randomly selected gadgets chain. It is clear that our GA-suggested ROP chain has minimal overhead in comparison to others. As the performance overhead is 2%, we claim this technique to be lightweight. Fig. 3.7 shows that the performance overhead generated by our security framework is negligible.

3.5.3 Attack detection

Fig. 3.8 shows how ROP chains help to detect whether a code reuse attack has been mounted or not. The ROP saves the return address of functions in a data segment and compares it with the return address of the calling function. If the addresses match then it allows the function to run, otherwise the binary should halt.

Table 3.1 Performance overhead

Technique used	Author	Overhead %
Lightweight CFI	Proposed method	2%
Fine-CFI	Li et al. [19]	10%
Parallax	Andriesse et al. [18]	4%
Practical CS-CFI	van der Veen et al. [23]	8.5%
CFI for COTS binaries	Zhang and Sekar [25]	8.5%

Table 3.2 The gadget execution time

Gadgets	Execution count	% of time it occurs in ROP chain
pop ecx; ret	60	0.60
pop edx; ret	58	0.58
pop eax; ret	57	0.57
pop eax; ret; Mov edx, eax; ret	54	0.54
pop ebx, pop esi; pop edi; ret	52	0.52
pop esi; ret	49	0.49
pop edi; ret	47	0.47
pop ebx; ret	47	0.47

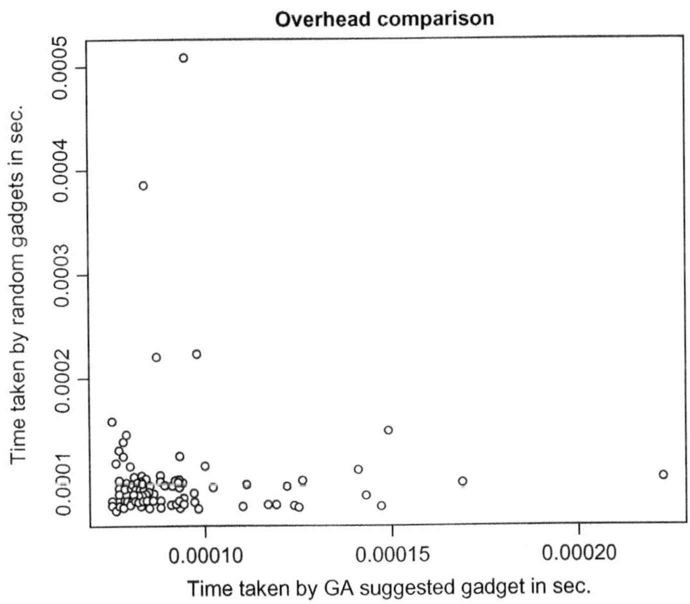

FIG. 3.7

Comparison of selected and random ROP chain.

As shown in Fig. 3.9 the ROP chain is stored in the stack. The top of the stack contains the gadget, which returns the offset value of the data; it executes and moves to the next gadget and continues up to the end of the stack.

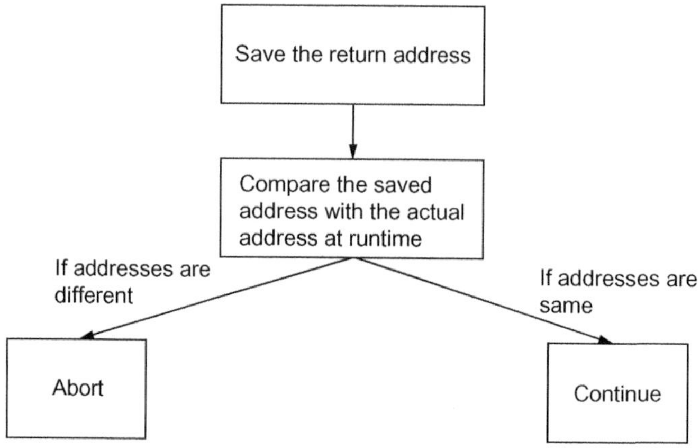

FIG. 3.8

Protection by using ROP.

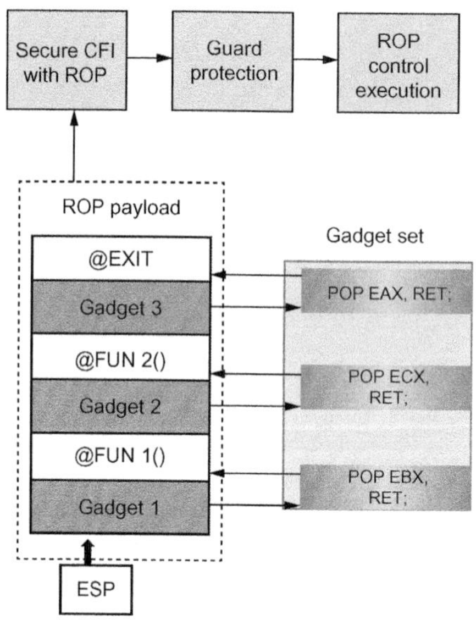

FIG. 3.9

Working ROP chain.

3.6 Conclusion

This chapter presented our proposal for developing a lightweight method for code self-verification of IoT devices. We developed a lightweight code self-integrity measure in the form of a dual implementation of ROP and code checksum. The strength of this security measure lies in the fact that these two security modules (ROP and code checksum) are not acting in isolation from one another but rather they are acting in tandem, complimenting each other. Therefore the only way to tamper with the code is to bypass both security layers at the same time. As such, the time and effort to bypass both layers increases the cost of attack, that is, it alerts the administrator about the attack. The only limitation of this model is that it is dependent on the programming language. Without monitoring there is a chance that an adversary can break the security framework. In the future, we plan to deploy a monitoring model that will help us to continuously monitor ROP chain execution without generating overhead to the main program.

Acknowledgments

This work is supported by the Department of Electronics and Information Technology (DeitY), Government of India, and the Netherlands Organization for Scientific research (NWO), Netherlands.

References

[1] B.S.P. Mishra, H. Das, S. Dehuri, A.K. Jagadev, Cloud Computing for Optimization: Foundations, Applications, and Challenges, Springer, Cham, 2018.

[2] B.B. Mishra, S. Dehuri, B.K. Panigrahi, A.K. Nayak, B.S.P. Mishra, H. Das, Computational Intelligence in Sensor Networks, Springer, Switzerland, 2019.

[3] H. Das, A.K. Jena, J.C. Badajena, C. Pradhan, R. Barik, Resource allocation in cooperative cloud environments, in: Progress in Computing, Analytics and Networking, Springer, Switzerland, 2018, pp. 825–841.

[4] P.K. Pattnaik, S.S. Rautaray, H. Das, J. Nayak (Eds.), Progress in Computing, Analytics and Networking: Proceedings of ICCAN 2017, vol. 710, Springer, Switzerland, 2018.

[5] K.H.K. Reddy, H. Das, D.S. Roy, A data aware scheme for scheduling big data applications with savanna hadoop, in: Networks of the Future*: Architectures, Technologies, and Implementations*, Chapman and Hall/CRC, New York, 2017, pp. 377–392.

[6] J. Nayak, B. Naik, A. Jena, R.K. Barik, H. Das, Nature inspired optimizations in cloud computing: applications and challenges, in: Cloud Computing for Optimization: Foundations, Applications, and Challenges, Springer, Cham, 2018, pp. 1–26.

[7] A. Akhunzada, M. Sookhak, N.B. Anuar, A. Gani, E. Ahmed, M. Shiraz, S. Furnell, A. Hayat, M.K. Khan, Man-at-the-end attacks: analysis, taxonomy, human aspects, motivation and future directions, J. Netw. Comput. Appl. 48 (2015) 44–57.

[8] H. Shacham, The geometry of innocent flesh on the bone: return-into-libc without function calls (on the x86), in: Proceedings of the 14th ACM Conference on Computer and Communications Security, ACM, 2007, pp. 552–561.

[9] H. Chang, M.J. Atallah, Protecting software code by guards, in: Security and Privacy in Digital Rights Management, Springer, 2002, pp. 160–175.

[10] Y. Chen, R. Venkatesan, M. Cary, R. Pang, S. Sinha, M.H. Jakubowski, Oblivious hashing: a stealthy software integrity verification primitive, in: International Workshop on Information Hiding, Springer, 2002, pp. 400–414.

[11] C.S. Collberg, C. Thomborson, Watermarking, tamper-proofing, and obfuscation-tools for software protection, IEEE Trans. Softw. Eng. 28 (8) (2002) 735–746.

[12] M. Abadi, M. Budiu, U. Erlingsson, J. Ligatti, Control-flow integrity, in: Proceedings of the 12th ACM Conference on Computer and Communications Security, ACM, 2005, pp. 340–353.

[13] G. Wurster, P.C. Van Oorschot, A. Somayaji, A generic attack on checksumming-based software tamper resistance, in: Security and Privacy, IEEE Symposium on, IEEE, 2005, pp. 127–138.

[14] M. Abadi, M. Budiu, Ú. Erlingsson, J. Ligatti, Control-flow integrity principles, implementations, and applications, ACM Trans. Inf. Syst. Secur. 13 (1) (2009) 4.

[15] A.F. Glew, D.A. Gerrity, C.T. Tegreene, Control Flow Integrity, US Patent App. 13/136,024 (January 24), 2013.

[16] C. Bhatt, N. Dey, A.S. Ashour, Internet of Things and Big Data Technologies for Next Generation Healthcare, Springer, Cham, 2017.

[17] N. Dey, A.S. Ashour, C. Bhatt, Internet of things driven connected health-care, in: Internet of Things and Big Data Technologies for Next Generation Healthcare, Springer, Cham, 2017, pp. 3–12.

[18] D. Andriesse, H. Bos, A. Slowinska, Parallax: implicit code integrity verification using return-oriented programming, in: Dependable Systems and Networks (DSN), 2015 45th Annual IEEE/IFIP International Conference on, IEEE, 2015, pp. 125–135.

[19] J. Li, X. Tong, F. Zhang, J. Ma, Fine-cfi: fine-grained control-flow integrity for operating system kernels, IEEE Trans. Inf. Forensics Secur. 13 (6) (2018) 1535–1550.

[20] N. Sasirekha, M. Hemalatha, A survey on software protection techniques against various attacks, Global J. Comp. Sci. Technol. 12 (1), 2012, 53–58, Version 1.0.

[21] X. Zhang, F. He, W. Zuo, Hash function based software watermarking, in: Advanced Software Engineering and Its Applications, 2008, ASEA 2008, IEEE, 2008, pp. 95–98.

[22] M. Jacob, M.H. Jakubowski, R. Venkatesan, Towards integral binary execution: implementing oblivious hashing using overlapped instruction encodings, in: Proceedings of the 9th Workshop on Multimedia & Security, ACM, 2007, pp. 129–140.

[23] V. van der Veen, D. Andriesse, E. Goktas, B. Gras, L. Sambuc, A. Slowinska, H. Bos, C. Giuffrida, Practical context-sensitive cfi, in: Proceedings of the 22nd ACM SIGSAC Conference on Computer and Communications Security, ACM, 2015, pp. 927–940.

[24] C. Zhang, T. Wei, Z. Chen, L. Duan, L. Szekeres, S. McCamant, D. Song, W. Zou, Practical control flow integrity and randomization for binary executables, in: Security and Privacy (SP), 2013 IEEE Symposium on, IEEE, 2013, pp. 559–573.

[25] M. Zhang, R. Sekar, Control flow integrity for cots binaries, in: USENIX Security Symposium, 2013, , pp. 337–352.

[26] R. Shrivastava, C. Hota, P. Shrivastava, Protection against code exploitation using ROP and check-summing in IoT environment, in: 2017 5th International Conference on Information and Communication Technology (ICoIC7) (ICoICT 2017), Melaka, Malaysia, 2017.

[27] C. Hota, R. Shrivastava, S. Shipra, Tamper-resistant code using optimal ROP gadgets for IoT devices, in: IWCMC 2017 M2M&IoT Symposium (IWCMC 2017 M2M&IoT Symposium), Valencia, Spain, 2017.

[28] E. Buchanan, R. Roemer, H. Shacham, S. Savage, When good instructions go bad: generalizing return-oriented programming to RISC, in: Proceedings of the 15th ACM Conference on Computer and Communications Security, ACM, 2008, pp. 27–38.

[29] S. Checkoway, L. Davi, A. Dmitrienko, A.-R. Sadeghi, H. Shacham, M. Winandy, Return-oriented programming without returns, in: Proceedings of the 17th ACM Conference on Computer and Communications Security, ACM, 2010, pp. 559–572.
[30] M. Prandini, M. Ramilli, Return-oriented programming, IEEE Secur. Priv. 10 (6) (2012) 84–87.
[31] A. Homescu, M. Stewart, P. Larsen, S. Brunthaler, M. Franz, Microgadgets: size does matter in turing-complete return-oriented programming, in: Proceedings of the 6th USENIX conference on Offensive Technologies, USENIX Association, 2012, p. 7.
[32] G. Elhayatmy, N. Dey, A.S. Ashour, Internet of things based wireless body area network in healthcare, in: Internet of Things and Big Data Analytics Toward Next-Generation Intelligence, Springer, 2018, pp. 3–20.

CHAPTER

Monte-Carlo Simulation models for reliability analysis of low-cost IoT communication networks in smart grid

4

Bhargav Appasani*, Dusmanta Kumar Mohanta†

*School of Electronics Engineering, KIIT University, Bhubaneswar, India †Department of Electrical and Electronics Engineering, Birla Institute of Technology, Mesra, India

4.1 Introduction

4.1.1 Introduction to the research work

In the last few decades, rapid progress was made in the fields of electronics and integrated circuit (IC) fabrication, which indirectly bolstered the growth of other technologies [1]. The Internet of things (IoT) is an emerging technology that is gradually proliferating into the fields of medicine, communication, sensor networks, transportation, and others. Its immense popularity and profound applications led researchers to incorporating IoT networks into smart grid (SG) technology [2, 3]. The SG is the next version of the existing power grid where the devices not only exchange energy but also exchange data [4]. The SG consists of phasor measurement units (PMUs) positioned at remote locations on the electrical buses communicating with the control center known as the phasor data concentrator (PDC) [5]. The sampling of data is time synchronized using the global positioning system (GPS) for calculating time-tagged voltage and current phasors. Thus real-time informatics pertaining to dynamics of the whole power industry can be obtained from the voltage and current phasors represented by respective magnitudes and phase angles, thereby providing adequate situational awareness for proactive measures to guard against different contingencies. The efficacy of the PMUs is proved by their ability to provide enhanced situational awareness to avoid catastrophic failures, such as blackouts, and has several other applications [6, 7].

Real-Time Data Analytics for Large Scale Sensor Data. https://doi.org/10.1016/B978-0-12-818014-3.00004-8

4.1.2 Motivation for the research and its objectives

The communication of phasor data from each PMU to the PDC can also be achieved using IoT networks. These networks should be reliable and cost-effective. The motivation for this work is to analyze the reliability of low cost IoT communication networks for PMUs in an SG. There are three major objectives of the research work as follows:

1. Design of wireless IoT networks for synchrophasor data communication.
2. Development of models for reliability assessment of these IoT communication networks.
3. Optimal placement of these networks for reducing the overall cost of the communication system for practical power grids of India.

4.1.3 Literature review

We conducted an extensive review of the communication technologies, architectures, and protocols for major SG applications [8–13]. With the sustained growth of wireless technology, wireless communication emerges as a viable alternative to the existing fiber optic-based communication [14] owing to several conspicuous advantages [15]. These include shorter installation time, shorter propagation delay, and lower maintenance cost, among others. Appasani and Mohanta demonstrated the advantages of wireless technology for practical synchrophasor applications [16]. The reliability of these networks was also calculated using fault tree diagrams. Similarly, series–parallel block diagrams and Markov models were proposed for estimating the reliability parameters [17, 18]. However, these approaches are simple to implement and cannot model simultaneous component failures nor can they model complexly interconnected systems. Also, they do not consider the uncertainties in failure rates and thus the reliability indices calculated using these approaches are less realistic. More complex methods such as Hidden Markov modeling [19], the system of systems engineering approach [20], dynamic fault tree approach [21], and others have been applied for reliability analysis of the PMUs but not for their communication networks. This work uses two-stage Monte-Carlo simulation (MCS) models for reliability analysis of wireless IoT communication networks.

4.1.4 Contributions

The main contribution of this chapter is the use of a two-stage MCS method for modeling the uncertainties in the failure rates of the components to provide a better estimate of the reliability parameters. The existing approaches presented in earlier works for estimating the reliability of PMU IoT networks cannot accommodate the uncertainty in the failure rates and considers them to be constant. Moreover, the series–parallel network approach makes it very difficult to calculate the mean time to failure (MTTF) and mean time to repair (MTTR) of IoT communication networks. An estimate of these parameters would be helpful in predicting the occurrence of failure and taking proper proactive measures.

The next major contribution of the chapter is the development of reliability models for PMU IoT communication networks. The previous works model the components as series–parallel networks thereby providing inaccurate results [16–18].

Another contribution of the work is the optimal placement of the PMUs such that the power system is observable at the expense of a minimum number of repeaters. Minimizing the repeaters reduces the cost of the communication networks and improves the availability of these networks. The efficacy of the proposed work for reliable synchrophasor data transferal has been validated using a representative practical case study with commensurate results for an Indian power grid.

4.1.5 Organization of the chapter

An overview of the wireless IoT communication networks for PMUs is provided in Section 4.2 to improve overall comprehension of the concepts introduced in this chapter. In order to proceed with the system reliability evaluation, the specifications of a practical communication network prevalent in communication parlance have been presented. Section 4.2 also discusses the basic concepts of reliability. Section 4.3 is dedicated to the description of the two-stage MCS method and its applicability to the present problem. Section 4.4 discusses the optimal placement of PMUs (OPP) for reducing the cost of communication networks without degrading the system's observability. Section 4.5 discusses the case study of four Eastern states in India and the final section summarizes the chapter and provides directions for future research.

4.2 Overview of wireless IoT communication networks for PMUs

4.2.1 Wireless communication networks

IEEE C37.118.2-2011 [22] specifies the standard for synchrophasor data transfer for power systems. According to this standard, the bandwidth requirement for a typical PMU data varies from 16 to 96 Kbps depending on the reporting rate, transmission protocol, the number of phasor measurements, and so on. This section presents a brief overview of the wireless IoT communication networks for the transfer of PMU data. Appasani and Mohanta present a complete description of these networks [16].

Wireless communication is a point-to-point communication, that is, communication takes place only when the line of sight exists between the transmitter and the receiver. Point-to-point wireless links are the most reliable and cost-efficient means of transferring high-speed data. An important component of this technology is the antenna. The length of the link depends on the size of the antenna, the height of the antenna, the frequency of operation, and the desired availability of the link. Proper link planning [23, 24] is essential for dependable transferal of the PMU data. To ensure a link availability of 0.999999 the maximum inter-tower distance between the successive repeaters has been taken as 25 km [25]. To improve the link availability, a twofold frequency diversity scheme has been proposed in earlier works and a similar redundancy is assumed for the proposed work. Furthermore the communication networks are assumed to be dedicated, that is, the communication resources are used only for the purpose of communication between PMUs and the PDC. The block diagram of the various components of the communication network is shown in Fig. 4.1 and the rate of failure and rate of repair for the remaining constituents are shown in Table 4.1.

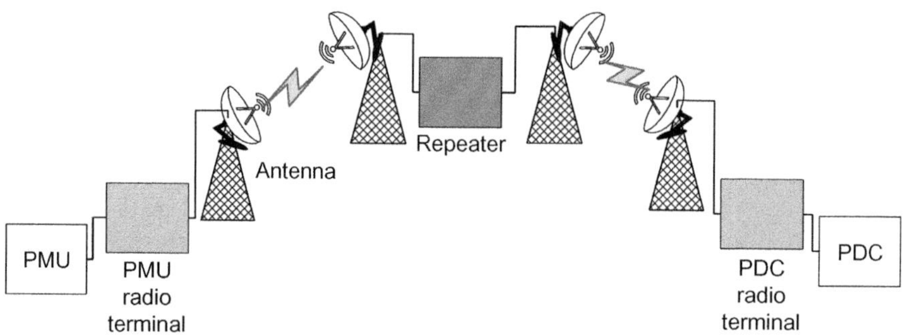

FIG. 4.1

Block diagram of a wireless PMU IoT communication network.

Table 4.1 Data corresponding to the various components of communication networks

No. (i)	Component	λ_i (failures/h)	μ_i (repair/h)
1	PMU radio terminal [26]	5.7078E−06	0.2222
2	Repeater [27]	1.1765E−05	0.2222
3	PDC radio terminal [26]	5.7078E−06	0.2222
4	Antenna [28]	3.47E−07	0.0417

To enhance the reliability, a redundant radio terminal is provided at the PMU and at the PDC. Also a redundant repeater is provided at each of the intermediate relaying towers. The radios at the PMU and those at the PDC operate in parallel, whereas the repeaters at each intermediate tower operate in hot standby mode. The interconnections among the various components of a communication network consisting of the single intermediate repeater are shown in Fig. 4.2.

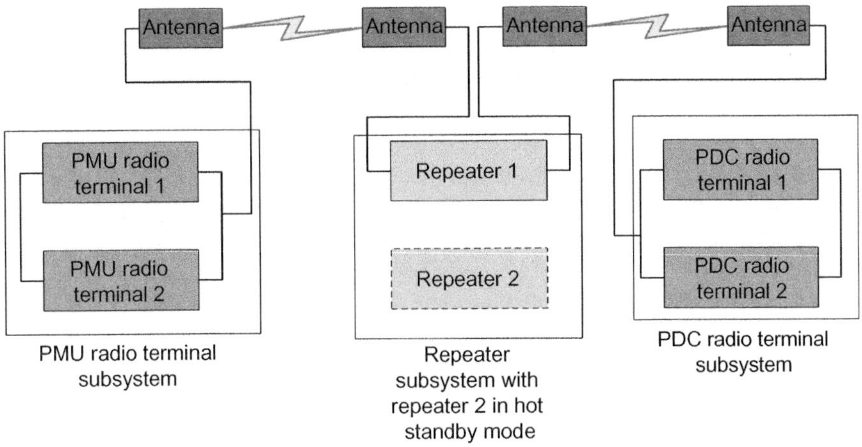

FIG. 4.2

Single repeater IoT communication network with component redundancy.

4.2.2 Reliability concepts

Reliability is an important concept in engineering system design [29]. Any potential technology or system should be evaluated for its reliability. Apart from being reliable, a system should also be evaluated for its maintainability and availability. Reliability is the probability that the system will perform its desired function over the specified time period. The reliability of a system with a failure rate of λ_s is given by Eq. (4.1).

$$R(t) = e^{-\lambda_s t} \tag{4.1}$$

Maintainability is the probability that a failed system could be restored back to its operating condition within the specified time period. The maintainability of a system having a repair rate of μ_s is given by Eq. (4.2).

$$H(t) = 1 - e^{-\mu_s t} \tag{4.2}$$

Availability is the probability that the system will perform its desired function at a given instance of time. The availability of the system is given by Eq. (4.3).

$$A = \left(\frac{\mu_s}{\mu_s + \lambda_s}\right) \tag{4.3}$$

4.3 Monte-Carlo simulation models

The failure of an IoT communication network is a rare event. However, from the practical failure rates of its constituents, it is feasible to estimate its failure statistics. Several approaches are available in the literature [16–21] with the most effective being the MCS. It empowers the engineer to emulate the real-life behavior of a system, thereby obtaining its performance statistics. Another reason for adopting this approach for the present scenario is the complexity of interactions that are involved in the communication network and the need for incorporating the uncertainties in the reliability analysis.

As described in Section 4.2, every communication network consists of three subsystems: the radio subsystem with its components operating in parallel, the repeater subsystem with the redundant component operating in hot standby mode, and the antenna subsystem. An IoT network with r intermediate towers consists of two radio terminal subsystems, one at the PDC and the other at the PMU. It also consists of r repeater subsystems and $2r+2$ antenna subsystems. In the first stage of the MCS each of these subsystems is modeled in order to obtain the MTTR and MTTF of each of these subsystems. These statistics are incorporated in the second stage to obtain the failure rates of the composite system.

For incorporating the uncertainties in the analysis, the MCS introduces random values of time to failure (TTF) and time to repair (TTR) at every iteration. These values are based on the nominal values of the component's failure rate and repair rate [18]. The random values for TTF and TTR used by the MCS are given by Eqs. (4.4) and (4.5).

$$\mathrm{TTF} = -\frac{1}{\lambda}\ln(\theta_1) \tag{4.4}$$

$$\mathrm{TTR} = -\frac{1}{\mu}\ln(\theta_2) \tag{4.5}$$

where θ_1 and θ_2 are random numbers between 0 and 1.

4.3.1 MCS model of the repeater subsystem

Each repeater subsystem consists of two repeaters operating in standby mode. The redundant component $R2$ comes into operation only when the main component $R1$ fails. The operation cycle of the two-component repeater subsystem is shown in Fig. 4.3.

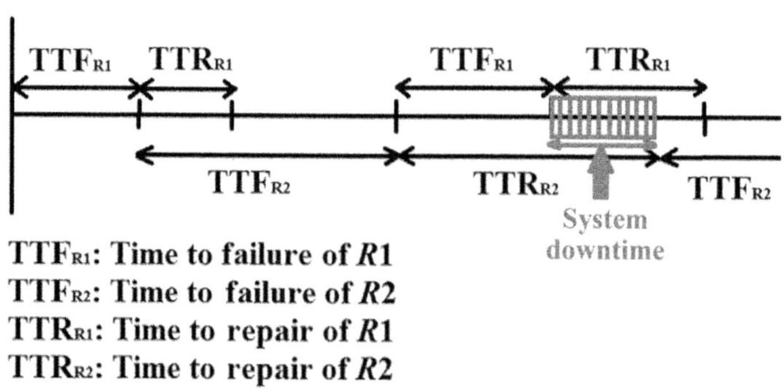

FIG. 4.3

Operation cycle for the repeater subsystem.

The subsystem fails when the standby component fails before the other component is repaired and is indicated by the red line (system downtime is the duration for which the system is unavailable) in Fig. 4.3. In order to generate sufficient failure data, the mission time *MT* should be taken sufficiently large. All the simulations performed in this work, consider a mission time of 1.752E+11 h. Each MCS is run for 1000 iterations in order to obtain consistent failure data. The flowchart for calculating the MTTF and MTTR of the repeater subsystem using the MCS is shown in Fig. 4.4 and the frequency distribution of failures for the repeater subsystem is shown in Fig. 4.5.

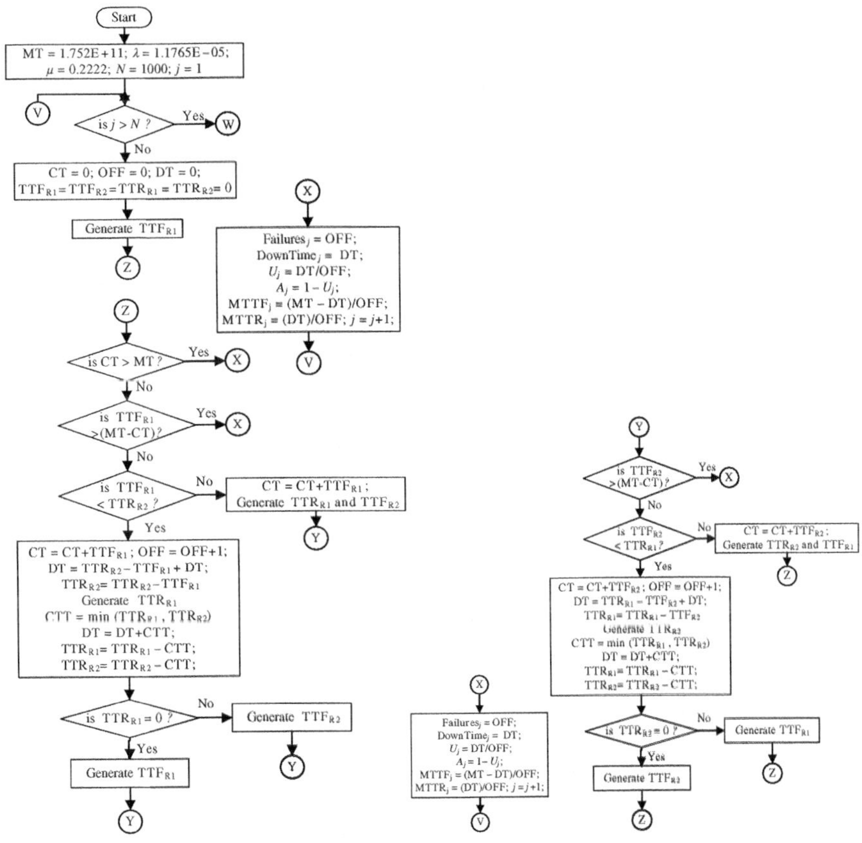

FIG. 4.4

MCS for repeater subsystem.

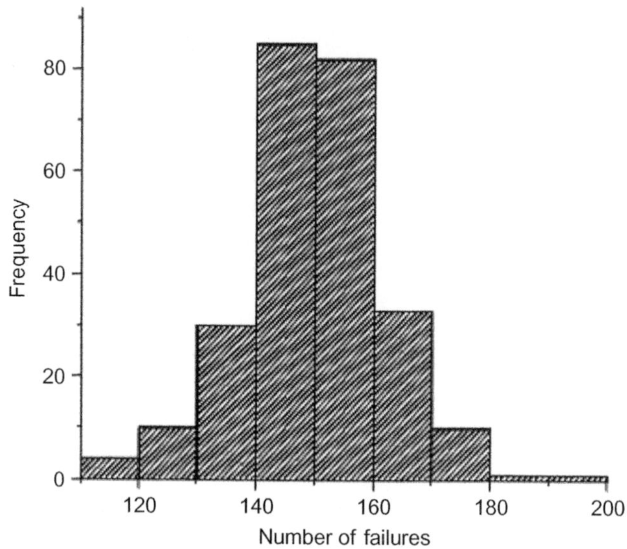

FIG. 4.5

Failure distribution for the repeater subsystem.

4.3.2 MCS model of the radio subsystem

Each radio subsystem consists of two terminals operating in parallel mode. The operation cycle of the two-component repeater subsystem is shown in Fig. 4.6.

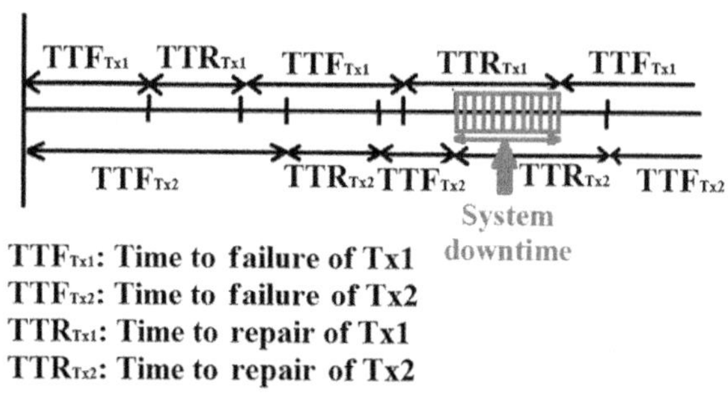

FIG. 4.6

Operation cycle for the radio subsystem.

The subsystem fails when both the components fail, and the subsystem downtime is indicated by the red lines in Fig. 4.6. The flowchart for calculating the MTTF and MTTR of the repeater subsystem using the MCS is shown in Fig. 4.7 and the frequency distribution of failures for the radio subsystem are shown in Fig. 4.8.

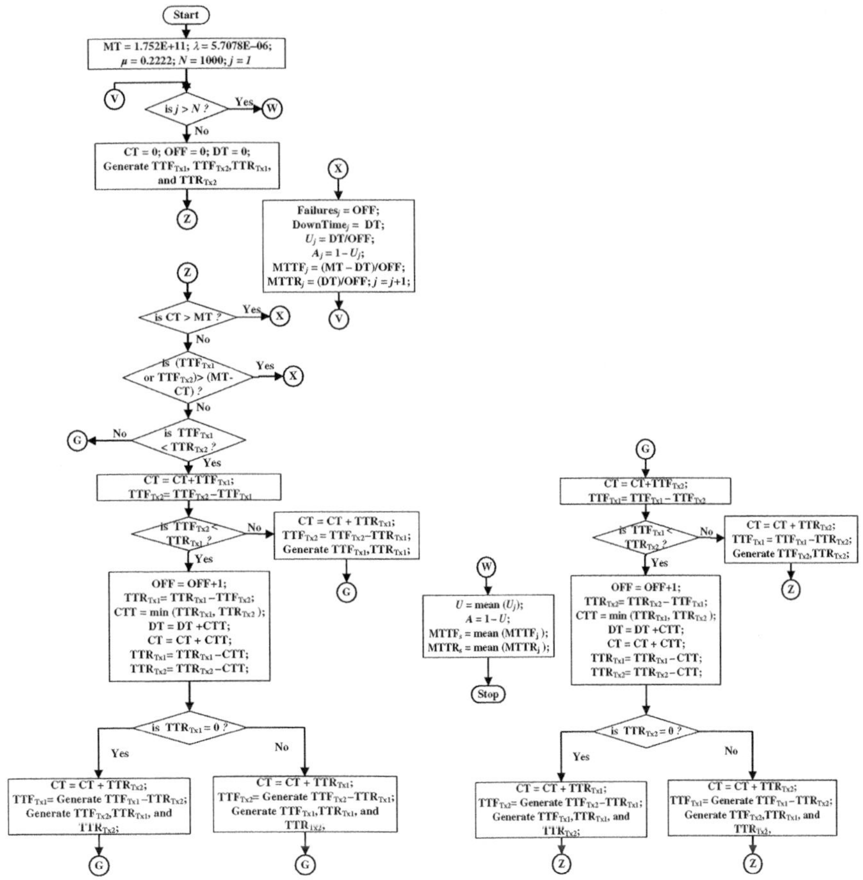

FIG. 4.7

MCS for radio subsystem.

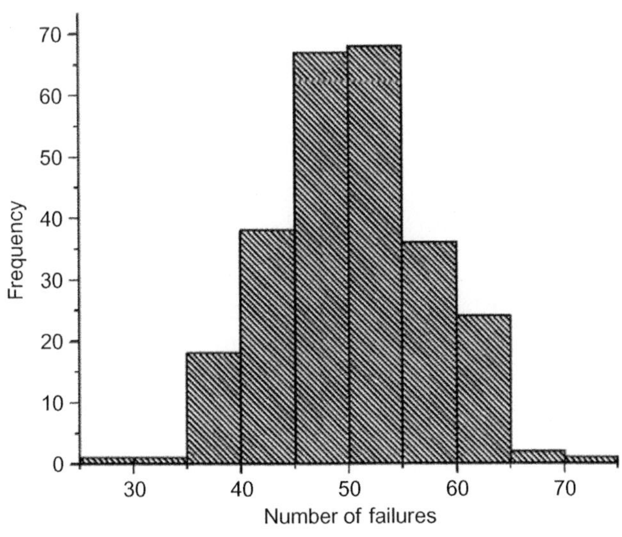

FIG. 4.8

Failure distribution for the radio subsystem.

4.3.3 MCS model of the antenna subsystem

The MCS model for the antenna subsystem is very simple as it does not have any redundant component. The subsystem fails when the antenna fails, and the subsystem downtime is same as the TTR of the antenna. The operation cycle is shown in Fig. 4.9 and the failure distribution is shown in Fig. 4.10. The rate of failure and rate of repair corresponding to that of the three subsystems are shown in Table 4.2.

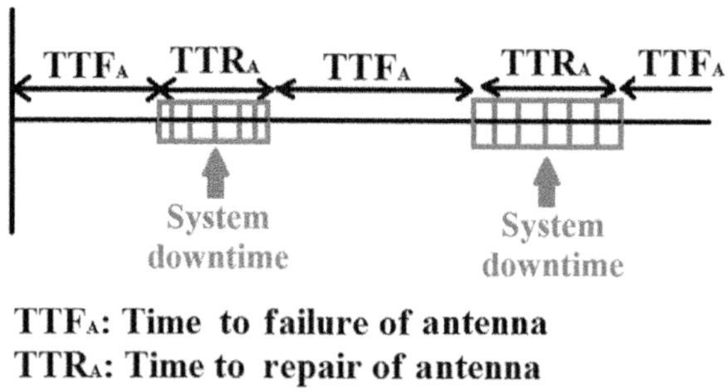

FIG. 4.9

Operation cycle for the antenna subsystem.

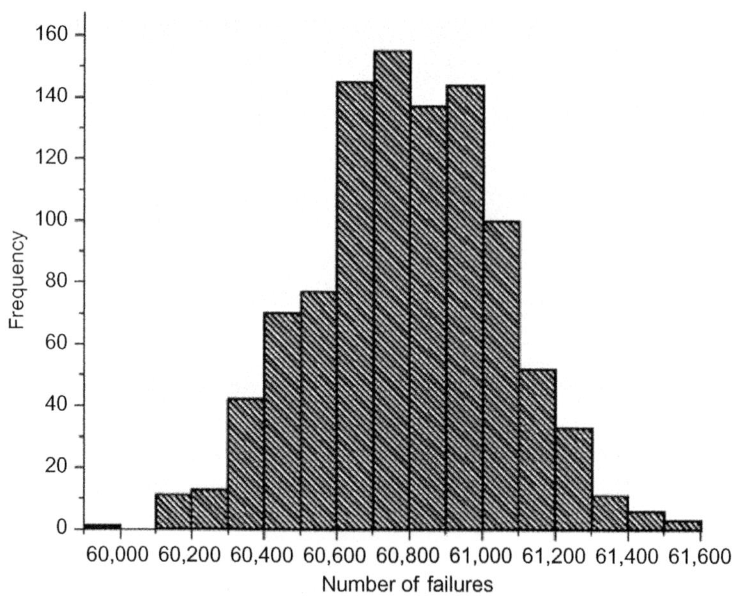

FIG. 4.10

Failure distribution for the antenna subsystem.

Table 4.2 MTTF and MTTR of the three subsystems

No.	Component	$MTTF_i$ (h)	$MTTR_i$ (h)	λ (failures/h)	μ (repair/h)
1.	Radio subsystem	34.923E+08	2.2333	2.8634E−10	0.4478
2.	Repeater subsystem	11.793E+08	7.1428	8.4796E−10	0.1400
3.	Antenna subsystem	2,881,844.3	24	3.47E−07	0.0417

From the failure distribution diagrams of all the three subsystems, it can be observed that the antenna subsystem is more prone to failure (due to lack of redundancy) and should be carefully monitored.

4.3.4 Second stage of MCS

After obtaining the failure rates and repair rates of the three subsystems, another MCS is performed to obtain the MTTF and MTTR of an r repeater IoT network. All the subsystems in the communication network are in series and the failure of any one of them leads to the failure of the entire IoT communication network. The flowchart for calculating the MTTF and MTTR of an r repeater IoT communication network is shown in Fig. 4.11 and their values are presented in Table 4.3.

4.4 Optimum PMU placement

PMU is a high-speed digital signal processor that is expensive and should be optimally used. Several techniques are available in the literature that deals with the optimum PMU placement (OPP) problem for total observability [30–35]. The concept behind the optimal placement of PMU is very simple. With the PMU installed on a bus, one can measure the voltage phasor of that bus and the current phasors of the incident electrical lines. Thus the voltage phasors of the connected buses can also be measured. When a PMU is placed on a bus in the power system, along with that bus all the adjacent buses are also observable [33]. The problem is to observe the power network with the smallest number of PMUs. To solve this combinatorial optimization problem, we use integer linear programming [34, 35] and graph theory. A network consisting of n nodes and b branches can be characterized using a connectivity matrix P of the order (n x n) as given by Eq. (4.6).

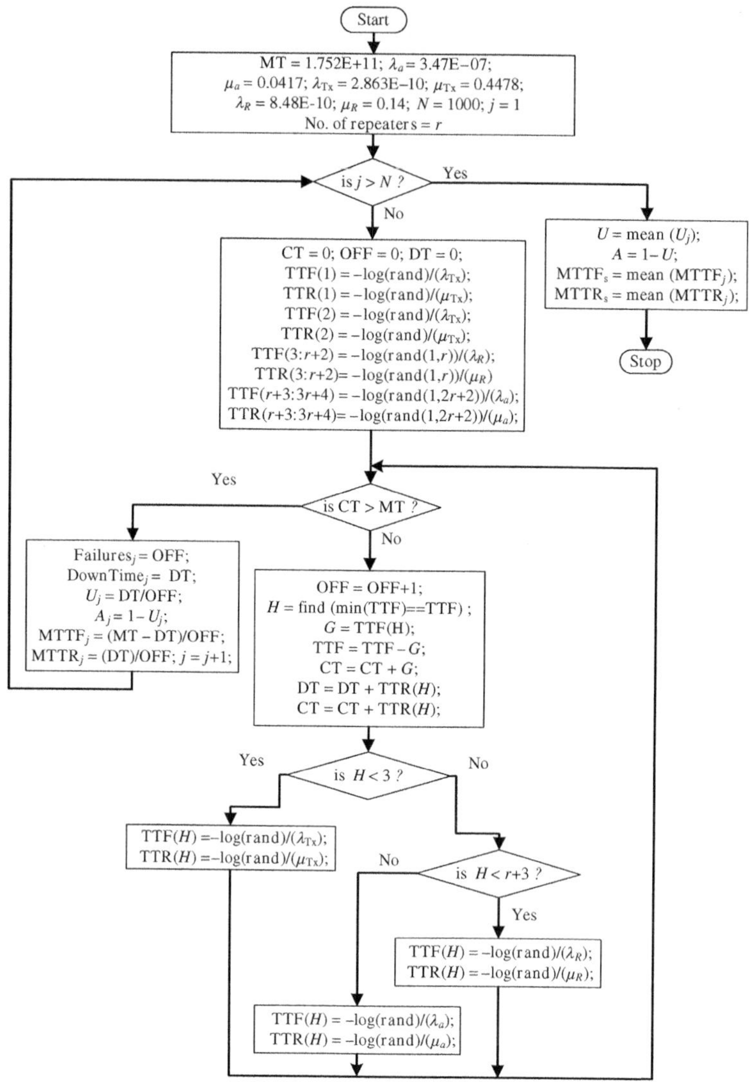

FIG. 4.11

Second stage MCS.

Table 4.3 MTTF and MTTR for an *r* repeater network

No. of repeaters *r*	$MTTF_s$ (h)	$MTTR_s$ (h)
0	1,439,774.3424	23.9808
1	719,776.8036	23.9832
2	479,812.0713	23.9810
3	359,818.5593	23.9800
4	287,863.4238	23.9797

$$p_{ij} = \begin{cases} 1 \text{ if } i\text{th bus is connected to } j\text{th bus or if } i = j \\ 0 \text{ elsewhere} \end{cases} \tag{4.6}$$

The location of the PMUs in the network is given by an $(n \times 1)$ location matrix X indicated by Eq. (4.7).

$$x_i = \begin{cases} 1 \text{ if the PMU is placed on the } i\text{th bus} \\ 0 \text{ elsewhere} \end{cases} \tag{4.7}$$

Using the ILP, the OPP problem is framed using Eq. (4.8), which can be easily solved using MATLAB.

$$N_{\min} = \min\left(\sum_{i=1}^{n} x_i\right) \tag{4.8}$$

With the constraint:

$$AX \geq [1\,1\ldots 1]^T_{(n\times 1)}.$$

where $N_{\min}$ is the least number of PMUs required for observing the power system.

After determining the optimal locations of the PMU, the next step is to calculate the repeaters that are needed by the PMU IoT communication networks. Depending on the choice of PDC, the repeaters needed for dependable communication of the PMU data will vary. The optimum locations for placement of PMUs and the PDC are those for which the number of repeaters is minimum and their locations are calculated using the digital elevation model data based approach [16–18].

The line of sight distance that the signal travels depends on the height of the towers. The propagation delay increases proportionally with the LoS distance. So it is essential that we optimize the tower heights. The well-known optimization algorithm, the genetic algorithm (GA) [36] has been used for optimizing the heights in order to reduce the propagation latency. The simulations were performed in MATLAB and the parameters of the GA are: number of generations is 100 with two-point crossover and mutation rate of 0.4. The fitness function considered for optimization is given by Eq. (4.9) [16].

$$f = \sum_{i=1}^{R-1}\left(\sqrt{D_i^2 + ((\text{alt}_{i+1} + H_{i+1}) - (\text{alt}_i + H_i))^2}\right) \tag{4.9}$$

where R is the number of nodes, D_i is the distance between the ith and $(i + 1)$th node in meters, alt_i is the altitude of the geographical location at the ith node in meters, and H_i is the height of the antenna tower for the ith node in meters. The first node is the PMU and the Rth node is the PDC. The flow chart for optimal placement of PMUs and the PDC for achieving total system observability with the minimum number of microwave towers is shown in Fig. 4.12.

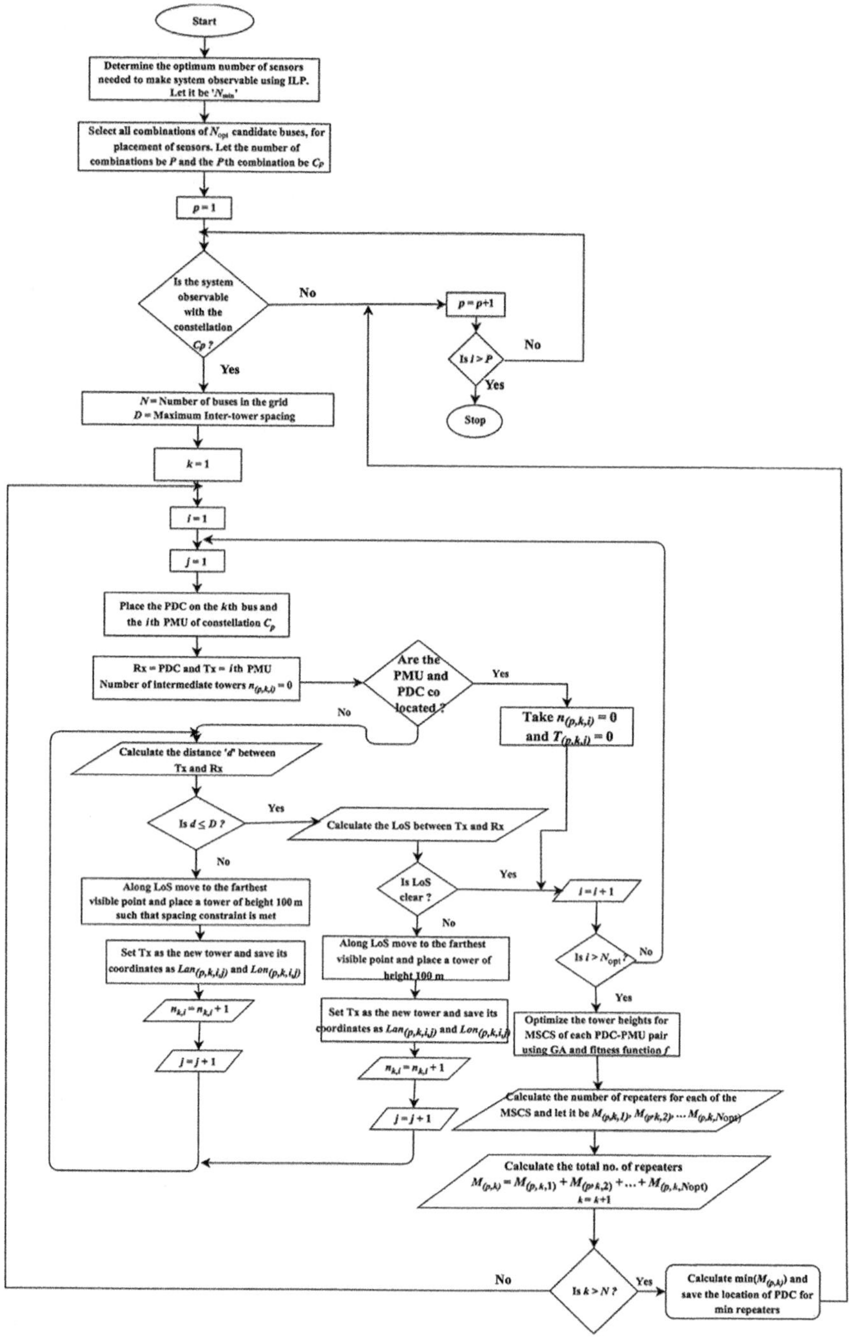

FIG. 4.12

Flowchart of the proposed algorithm based on GTOPO30 DEM data for the OPP.

4.5 Case study

Low-cost IoT communication networks were designed for the Eastern power grid of India. This power grid consists of four states: Bihar, Odisha, West Bengal, and Jharkhand. The simulation results for Odisha's power grid are provided in detail and only a summary of the results for the remaining power grids is given. The single line diagram for Odisha's grid is shown in Fig. 4.13 and its geographical data are given in Table 4.4 [16–18].

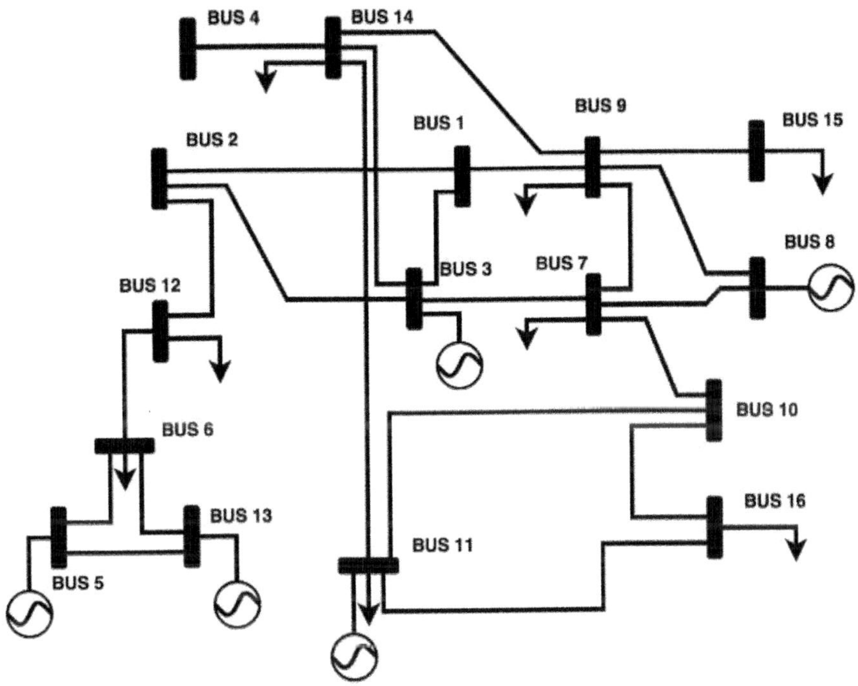

FIG. 4.13

Odisha's power grid.

Table 4.4 Geographical data of power grid of Odisha

Bus	Location	Latitude (°N)	Longitude (°E)
1	Rengali-I	20.4341	84.5845
2	Indravati	18.7418	80.2614
3	Talcher	20.5644	85.1326
4	Rourkela	22.1357	84.4832
5	Balimela	18.2244	82.0598
6	Jeypore	18.5219	82.3413

Continued

Table 4.4 Geographical data of power grid of Odisha *Continued*

Bus	Location	Latitude (°N)	Longitude (°E)
7	Talcher	21.0459	85.0823
8	Nalco	20.8653	85.1842
9	Rengali-II	20.4001	84.5049
10	Meramundali	20.6102	85.201
11	Bhanjanagar	19.9358	84.5825
12	Jeypore	18.5839	82.3240
13	Upper Kolab	18.8032	82.6022
14	Rourkela	22.1517	84.3847
15	Daitari	21.111	85.8255
16	Bhubaneswar	20.2961	85.8245

The optimum locations for PMUs and the PDC are obtained using the proposed algorithm. These locations are: Bus 1 (Rengali-I), Bus 6 (Jeypore), Bus 9 (Rengali-II), Bus 10 (Meramundali), and Bus 14 (Rourkela) for the PMUs, and Bus 1 (Rengali-I) for the PDC, with a total of 30 repeaters required for communication by the IoT networks. The locations for these repeaters along with the optimized tower heights are shown in Table 4.5 and the various elements of the IoT communication network between the PMU at Bus 10 and the PDC at bus 1 are shown in the Fig. 4.14.

Using the two-stage MCS, the MTTF and MTTR for each of the IoT networks (between every PMU-PDC pair) has been obtained and is also shown in Table 4.5. Also shown in the table is the various system parameters evaluated using Eqs. (4.3)–(4.5). The frequency of failures of these communication networks is shown in Fig. 4.15. These results are also compared with those obtained using the existing methods and the comparison is shown in Table 4.6.

From the reliability analysis of the power grid of Odisha and the failure distribution diagram, it can be observed that the IoT communication network between Jeypore and Rengali-I is more prone to failure and should be carefully monitored. Another observation from the failure distribution diagram is that the IoT communication network with a higher number of repeaters has a higher probability of failure and therefore minimizing the number of repeaters would not only reduce the cost but also enhance the availability of these networks. Also, from Table 4.6 it can be observed that the series–parallel method [17] and the fault tree method [16] cannot be used for determining the parameters such as the MTTF and MTTR. The Markov model can estimate these parameters but the values are less accurate than those obtained using the proposed method as it can model multiple failures simultaneously. Furthermore the proposed algorithm was applied to determine the optimal locations for placement of PMUs and the PDC for the remaining states, namely, Jharkhand, West Bengal, and Orissa. The reliability parameters evaluated for the constituent IoT communication networks of these states, using the two-stage MCS, is shown in Table 4.7.

Table 4.5 Geographical coordinates of the repeaters for power grid of Odisha

	IoT network between PMU at Bus 6 and PDC at Bus 1			IoT network between PMU at Bus 9 and PDC at Bus 1			IoT network between PMU at Bus 10 and PDC at Bus 1			IoT network between PMU at Bus 14 and PDC at Bus 1		
	Latitude °N	**Longitude °E**	**Tower height (m)**	**Latitude °N**	**Longitude °E**	**Tower height (m)**	**Latitude °N**	**Longitude °E**	**Tower height (m)**	**Latitude °N**	**Longitude °E**	**Tower height (m)**
PMU Radio	18.5219	82.3413	96.67	20.4001	84.5049	51.62	20.6102	85.201	98.38	22.1517	84.3847	99.21
Repeater 1	18.5668	82.3930	61.99	–	–	–	20.5701	85.0597	51.80	21.9575	84.4075	69.01
Repeater 2	18.7079	82.5559	52.22	–	–	–	20.5065	84.8369	98.22	21.7410	84.4329	95.22
Repeater 3	18.8553	82.7264	96.42	–	–	–	20.4618	84.6810	51.73	21.5916	84.4504	97.36
Repeater 4	19.0025	82.8972	70.75	–	–	–	–	–	–	21.3900	84.4739	55.38
Repeater 5	19.1304	83.0460	51.60	–	–	–	–	–	–	21.1734	84.4991	97.62
Repeater 6	19.2385	83.1720	89.01	–	–	–	–	–	–	21.0764	84.5103	50.91
Repeater 7	19.3655	83.3205	50.06	–	–	–	–	–	–	20.8598	84.5350	99.88
Repeater 8	19.5115	83.4915	95.58	–	–	–	–	–	–	20.8000	84.5423	52.65
Repeater 9	19.6573	83.6628	57.79	–	–	–	–	–	–	20.5837	84.5673	99.24
Repeater 10	19.7712	83.7971	76.08	–	–	–	–	–	–	20.5312	84.5733	50.78
Repeater 11	19.9167	83.9690	61.16	–	–	–	–	–	–	20.4416	84.5836	57.04
Repeater 12	20.0416	84.1170	51.81	–	–	–	–	–	–	–	–	–
Repeater 13	20.1913	84.2948	95.71	–	–	–	–	–	–	–	–	–
Repeater 14	20.2723	84.3913	50.90	–	–	–	–	–	–	–	–	–
Repeater 15	20.4217	84.5696	99.65	–	–	–	–	–	–	–	–	–
Repeater 16	20.4279	84.5771	53.40	–	–	–	–	–	–	–	–	–
PDC Radio	20.4341	84.5845	97.54	20.4341	84.5845	99.84	20.4341	84.5845	99.76	20.4341	84.5845	98.41
No. of repeaters	16			0			3			11		
$MTTF_s$ (h)	84,659.5			1,439,774.3			359,818.6			119,937.5		
$MTTR_s$ (h)	23.98			23.9808			23.98			23.9802		
R (1000h)	0.9883			0.9993			0.9972			0.9917		
H (100h)	0.984550			0.984547			0.984550			0.984549		
A	0.99971683			0.99998334			0.99993336			0.99980010		
Propagation delay (ms)	1.0568			0.0304			0.2238			0.6404		

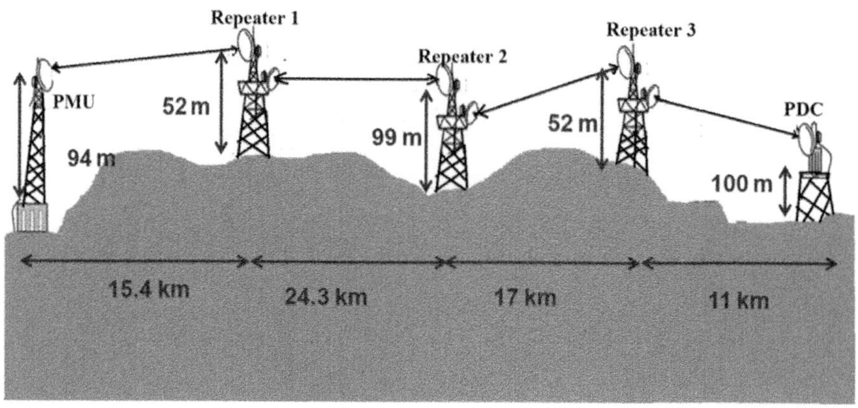

FIG. 4.14

Elements of IoT communication network between PMU at bus 10 and PDC at bus 1.

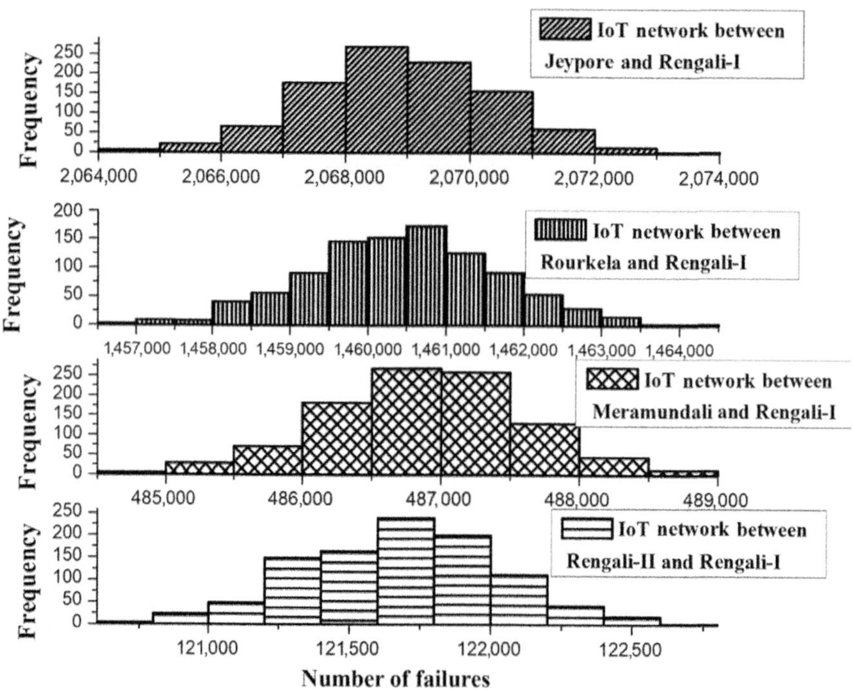

FIG. 4.15

Failure distribution for the power grid of Odisha.

Table 4.6 Comparison with the existing techniques

IoT communication network	Parameter	Proposed method	Markov model aided frequency-duration approach [18]	Series–parallel networks [17]	Fault tree analysis [16]
Between PMU at Bus 6 and PDC at Bus 1	Availability	0.99971683	0.999717069	0.999716826	0.999716826
	MTTF (h)	84,659.5	84,613	Difficult to determine	Difficult to determine
	MTTR (h)	23.98	23.95	Difficult to determine	Difficult to determine
Between PMU at Bus 9 and PDC at Bus 1	Availability	0.999983344	0.999983356	0.999983356	0.999983356
	MTTF (h)	1,439,774.3	1,439,705.5	Difficult to determine	Difficult to determine
	MTTR (h)	23.98	23.96	Difficult to determine	Difficult to determine
IoT network between PMU at Bus 10 and PDC at Bus 1	Availability	0.999933360	0.999933372	0.999933372	0.999933372
	MTTF (h)	359,818.6	359,670.3	Difficult to determine	Difficult to determine
	MTTR (h)	23.98	23.95	Difficult to determine	Difficult to determine
Between PMU at Bus 14 and PDC at Bus 1	Availability	0.99980010	0.99980028	0.99980011	0.99980011
	MTTF (h)	119,937.5	119,871.1	Difficult to determine	Difficult to determine
	MTTR (h)	23.98	23.95	Difficult to determine	Difficult to determine

Table 4.7 Reliability parameters evaluated for the power grid of Jharkhand, Bihar, and West Bengal

State	Location of PMUs	No. of repeaters	PDC	$MTTF_s$ (h)	$MTTR_s$ (h)	*R* (1000 h.)	*H* (100 h.)	*A*	Propagation delay (ms)
Jharkhand	Tenughat 23.4010°N 85.5222°E	–	Tenughat 23.4010°N 85.5222°E	–	–	–	–	–	–
	Patratu 23.3231°N 85.2951°E	0		1,439,774.3	23.9808	0.9993	0.984547	0.99998334	0.0825
	Bokaro 23.4708°N 85.5829°E	0		1,439,774.3	23.9808	0.9993	0.984547	0.99998334	0.0331
	Jamshedpur 22.5417°N 86.0902°E	4		287,863.4	23.9797	0.9965	0.984551	0.99991670	0.3789
	Maithon 23.4630°N 86.4824°E	4		287,863.4	23.9797	0.9965	0.984551	0.99991670	0.3334
Bihar	Begusarai 25.2446°N 86.0734°E	3	Biharshariff 25.2709°N 85.3206°E	359,818.6	23.98	0.9972	0.984550	0.99993336	0.2525
	Fatuah 25.3105°N 85.1911°E	0		1,439,774.3	23.9808	0.9993	0.984547	0.99998334	0.0458
	Bihar Shariff 25.2218°N 85.3138°E	0		1,439,774.3	23.9808	0.9993	0.984547	0.99998334	0.0183
	Gaya 24.4758°N 85.0055°E	3		359,818.6	23.98	0.9972	0.984550	0.99993336	0.3132

West Bengal	Farakka 24.4032°N 87.5543°E	9	Waria 22.2359°N 88.1304°E	143,926.6	23.9806	0.9931	0.984548	0.99983341	0.8404
	Kolaghat 22.2639°N 87.5202°E	2		479,812.1	23.9810	0.9979	0.984547	0.99995002	0.2126
	Jeerat 22.4458°N 88.323°E	1		719,776.8	23.9832	0.9986	0.984545	0.99996668	0.1035
	Arambagh 22.501°N 87.5734°E	2		479,812.1	23.9810	0.9979	0.984547	0.99995002	0.2126
	Waria 22.2359°N 88.1304°E	–		–	–	–	–	–	–
	Lakshmikantpur 22.1099°N 88.3209°E	0		1,439,774.3	23.9808	0.9993	0.984547	0.99998334	0.0804
	Dalkhola 25.5101°N 87.51°E	15		89,948.2	23.9799	0.9889	0.984550	0.99973347	1.2542

4.6 Conclusion and directions for future research

The reliability of the synchrophasor communication system for the power industry has been a challenge, to ensure its real-time monitoring and control. The limitations of fiber optic-based communication mostly arise due to harsh terrains as well as increased cost for longer distances. The detection and localization of fault are also cumbersome as well as time-consuming and thus adversely affect reliability. The various system parameters such as availability, reliability, maintainability, and quality of service for the IoT communication networks have been explored using a two-stage MCS for the four Eastern states of India to corroborate the effectiveness of the methodology. Security in wireless IoT communication networks is still a major challenge. Another important aspect is the integration of cloud computing and big data analytics along with IoT for SG applications [37, 38]. The authors intend to investigate this in their future research work.

References

[1] J.A. Stankovic, Research directions for the internet of things, IEEE Internet Things J. 1 (1) (2014) 3–9.

[2] L. Atzori, A. Iera, G. Morabito, The internet of things: a survey, Comput. Netw. 54 (15) (2010) 2787–2805.

[3] S. Chen, H. Xu, D. Liu, B. Hu, H. Wang, A vision of IoT: applications, challenges, and opportunities with China perspective, IEEE Internet Things J. 1 (4) (2014) 349–359.

[4] Y. Li, X. Cheng, Y. Cao, D. Wang, L. Yang, Smart choice for the smart grid: narrowband internet of things (NB-IoT), IEEE Internet Things J. 5 (3) (2018) 1505–1515.

[5] J.D.L. Ree, V. Centeno, J.S. Thorp, A.G. Phadke, Synchronized phasor measurement applications in power system, IEEE Trans. Smart Grid 1 (1) (2010) 20–27.

[6] A.G. Phadke, J.S. Thorp, Phasor measurement units and phasor data concentrators, in: Synchronized Phasor Measurements and Its Applications, Springer, New York, 2008, pp. 93–105. Chap. 5.

[7] V.C. Gungor, et al., Smart grid technologies: communication technologies and standards, IEEE Trans. Ind. Inform. 7 (4) (Nov. 2011) 529–539.

[8] A. Avizienis, et al., Basic concepts and taxonomy of dependable and secure computing, IEEE Trans. Depend. Secure Comput. 1 (1) (2004) 11–33.

[9] A. G. Phadke and J. S. Thorp, "Communication needs for wide area measurement applications, " IEEE International Conference on Critical Infrastructure (CRIS), 2010.

[10] R.H. Khan, J.Y. Khan, A comprehensive review of the application characteristics and traffic requirements of a smart grid communications network, Comput. Netw. 57 (3) (2013) 825–845.

[11] M. Kuzlu, M. Pipattanasomporn, S. Rahman, Communication network requirements for major smart grid applications in HAN, NAN and WAN, Comput. Netw. 67 (4) (2014) 74–88.

[12] W. Wang, Y. Xu, M. Khanna, A survey on the communication architectures in smart grid, Comput. Netw. 55 (15) (2011) 3604–3629.

[13] B. Appasani, D.K. Mohanta, A review on synchrophasor communication system: communication technologies, standards and applications, Prot. Cont. Modern Power Syst. 3 (37) (2018) 1–17.
[14] R. Zurawski, From wireline to wireless networks and technologies, IEEE Trans. Ind. Inform. 3 (2) (May 2007) 93–94.
[15] T. Naveh, Mobile backhaul: Fiber vs. microwave, Ceragon White Paper, 1 pp. 1–11, 2009.
[16] B. Appasani, D.K. Mohanta, Co-optimal placement of PMUs and their communication infrastructure for minimization of propagation delay in WAMS, IEEE Trans. Ind. Inform. 14 (5) (2018) 2120–2132.
[17] B. Appasani, D.K. Mohanta, Optimal placement of synchrophasor sensors for risk hedging in a smart grid, IEEE Sensors J. 17 (23) (2017) 7857–7865.
[18] B. Appasani, D.K. Mohanta, A two-stage Markov model aided frequency duration approach for reliability analysis of PMU microwave communication networks, J. Risk Reliab. (2018).
[19] Y. Wang, W. Li, J. Lu, Reliability analysis of phasor measurement unit using hierarchical Markov modeling, Electr. Power Compon. Syst. 37 (5) (2009) 517–532.
[20] M. Cherukuri, D.S. Roy, D.K. Mohanta, Reliability evaluation of phasor measurement unit: a system of systems approach, Electr. Power. Compon. Syst. 43 (4) (2015) 437–448.
[21] P. Zhang, et al., Reliability evaluation of phasor measurement unit using Monte Carlo dynamic fault tree method, IEEE Trans. Smart Grid 3 (3) (2012) 1235–1243.
[22] IEEE Power & Energy Society IEEE standard for synchrophasor data transfer for power systems, IEEE Std C37.118.2-2011 (Revision of IEEE Std. C37.118-2005), 1–53, 2011.
[23] R.L. Freeman, Radio System Design for Telecommunication, Wiley, Hoboken, NJ, 1997.
[24] H. Lehpamer, Microwave Transmission Networks: Planning, Design and Deployment, Mc-Graw Hill, New York, 2004.
[25] Fcc.gov, '47CFR', 2016. [Online]. Available: https://www.fcc.gov/general/rules-regulations-title-47. Accessed: 22-Nov-2016.
[26] Harris.com 'TRuepoint 5000'2004. [Online]. Available: http://teleportpskov.ru/files/docs/harris/1_Datasheet_Truepoint.pdf. Accessed: 23-Aug-2016.
[27] Peninsulaengineering.com 'RF-6000EW Repeater' 2008. [Online]. Available: http://www.peninsulaengineering.com/pdfs/RF-6000EW_Brochure_C1.pdf. Accessed: 23-Aug-2016.
[28] Telecom-info.telcordia.com 'SR-332'2016. [Online]. Available: http://telecom-info.telcordia.com/site-cgi/ido/docs.cgi/sr 332. Accessed: 23-Aug-2016.
[29] C.E. Ebeling, An Introduction to Reliability and Maintainability Engineering, Mc-Graw Hill, New York, 2004.
[30] S. Chakrabarti, E. Kyriakides, Optimal placement of phasor measurement units for power system observability, IEEE Trans. Power Syst. (3) (2008) 1433–1440.
[31] K. K. More and H. T. Jadhav. "A literature review on optimal placement of phasor measurement units." IEEE International Conference on Power, Energy and Control (ICPEC), 2013.
[32] T.L. Baldwin, et al., Power system observability with minimal phasor measurement placement, IEEE Trans. Power Syst. 8 (2) (1993) 707–715.
[33] N.M. Manousakis, G.N. Korres, P.S. Georgilakis, Taxonomy of PMU placement methodologies, IEEE Trans. Power Syst. 27 (2) (2012) 1070–1077.

[34] B. Gou, Optimal placement of PMUs by integer linear programming, IEEE Trans. Power Syst. 283 (3) (Aug 2008) 1525–1526.
[35] B. Gou, Generalized integer linear programming formulation for optimal PMU placement, IEEE Trans. Power Syst. 23 (3) (2008) 1099–1104.
[36] D.E. Goldberg, Genetic Algorithms, Pearson Education, India, 2006.
[37] B.S.P. Mishra, H. Das, S. Dehuri, A.K. Jagadev, Cloud Computing for Optimization: Foundations, Applications, and Challenges, Springer, 2018.
[38] K.H.K. Reddy, H. Das, D.S. Roy, A data aware scheme for scheduling big data applications with SAVANNA Hadoop, in: Networks of the Future, Chapman and Hall, 2017, pp. 372–399.

CHAPTER

5 Lightweight ciphertext-policy attribute-based encryption scheme for data privacy and security in cloud-assisted IoT

Syam Kumar Pasupuleti*, **Dheerendra Varma**[†]
**IDRBT, Hyderabad, India* [†]*University of Hyderabad, Hyderabad, India*

5.1 Introduction

The Internet of things (IoT) is a network of physical devices that have unique IP addresses and the ability to sense and collect data from the external world. In IoT the devices exist in different geographic locations, and information collected through the devices can be used for different applications in areas such as healthcare, banking, and others [1]. Due to lack of memory, processing power, and bandwidth, IoT devices can't handle large and complex data. Therefore the cloud is the best option for storing this type of data as it provides scalable storage capacity under a pay-per-use model.

The cloud provides resources such as large storage space and processing capabilities at low costs to enable the use of real computing machines. The cloud is also useful for hosting websites and mobile applications for increasing resources to achieve scalability. However, moving data to the cloud brings two main concerns: (1) privacy and (2) access control [2]. For this reason, we need to ensure data privacy and provide access control for authorized users of cloud-assisted IoT.

The traditional solution to protect the privacy of data is encryption, but encryption schemes do not provide access control. To provide access control, various traditional approaches have been proposed, including mandatory access control (MAC), discretionary access control (DAC), and role-based access control (RBAC). In MAC, security labels are assigned to all the resources and users in the organization. Whenever a user wants to access the resources the operating system checks the user security label with the resource security label. If the two match, then resource access is allowed. If not, access is denied. The problem with this model is the amount of preplanning necessary before it can be effectively implemented. Management overhead for this model is high as security labels need frequent updating. In DAC, which is based on an access control list (ACL), users control their own data. A user gives the permission to access the system resources. The user's access

Real-Time Data Analytics for Large Scale Sensor Data. https://doi.org/10.1016/B978-0-12-818014-3.00005-X

permission details are stored in the ACL. DAC is not suitable for use when the data is stored in distributed environments. In RBAC, all users are assigned roles based on their job function in the organization. Resource access is based on these assigned roles. If we want to suspend a user, we have to suspend the entire group of users assigned the same roles. Therefore RBAC is not optimal because large groups would have the same type of access. Although all these methods provide access control, they do not provide data privacy.

Attribute-based encryption (ABE) overcomes the issues of traditional access control methods and provides data privacy as well [3]. There are two variations of ABE: key-policy attribute-based encryption (KP-ABE) and ciphertext-policy attribute-based encryption (CP-ABE). CP-ABE is the best approach for ensuring data privacy and fine-grained access control in the cloud environment because the data owner has control over the access policy.

In CP-ABE, data owners encrypt the data under the access policy. The data can be decrypted by users only if their attributes fulfill the owner-defined access policy. Recently several CP-ABE schemes were proposed based on different access structures. These include access trees [4–9], AND gates [10–13], and linear secret sharing schemes (LSSS) [3, 14–16]. However, all these existing CP-ABE schemes increase computation overhead for data encryption due to their access structure representation.

To make CP-ABE more efficient, we need to focus on the access structure [17]. These drawbacks motivated us to design a lightweight ciphertext-policy attribute based encrption (LCP-ABE) scheme with a new access structure. This chapter presents and discusses our LCP-ABE scheme for ensuring data privacy using cloud-assisted IoT.

In our LCP-ABE, we use a Walsh–Hadamard transform for data privacy and access. Our scheme supports efficient user revocation and is secure against collusion attacks, chosen-plaintext attacks (CPA), and forward secrecy (FS). It is also lightweight and more suitable for IoT devices.

In this chapter we describe related work (Section 5.2), discuss preliminaries (Section 5.3), define our system model (Section 5.4), present construction of the LCP-ABE scheme (Section 5.5), analyze the system's security (Section 5.6), and conclude our findings (Section 5.7).

5.2 Related work

Many researchers have proposed CP-ABE approaches based on different access structures to ensure data privacy and access control. As discussed in the previous section, there are three types of CP-ABE schemes: (1) access trees with threshold gates, (2) AND gates, and (3) linear secret sharing schemes (LSSS).

5.2.1 CP-ABE scheme with threshold gates

Bethencourt et al. [5] were first to design a CP-ABE scheme based on an access tree with threshold gates and encrypted message stores. In their scheme, decryption of the information can be done using Shamir's Secret Sharing [18]. Goyal et al. [6]

proposed a "bounded CP-ABE" that generalizes transformation from KP-ABE to CP-ABE through a universal access tree. This scheme accepts any access policies with bounded polynomial size but this scheme has a drawback that user is strictly use the depth of the access tree. Ibraimi et al. [9] designed a new method by using n-array access tree with AND and OR nodes, but the decryption operation does not work based on secret sharing. Guan et al. [7] designed a scheme in which the encryption algorithm divides the data into chunks and encrypts it in sequence with its corresponding subtree. Alrawais et al. [4] used the CP-ABE scheme for designing key exchanging and establishing a secure communication among the fog nodes. Hur and Noh [8] addressed the issues of attribute and user revocation. In their scheme, the service provider generates the key encryption keys (KEK) for each user after the setup phase is complete. The generated ciphertext is re-encrypted by using the attribute group key and this key is also encrypted with KEK. Hence unauthorized users are unable to perform the decryption operation. However, it is difficult to maintain all KEK for a large number of users in an organization.

Although threshold-based schemes achieve data privacy and access control, they are not suitable for lightweight IoT devices.

5.2.2 CP-ABE scheme with AND gates

AND gates follow the bit-wise representation of attribute presence in the access policy. Several researchers have proposed CP-ABE schemes based on AND gates. Cheung and Newport [10] described a new CP-ABE by using the AND gate access structure on "positive and negative attributes." Nishide et al. [12] proposed a CP-ABE scheme to hide the "access policy" with partially hidden access structures. Similarly, Emura et al. [11] proposed a scheme that enables a constant ciphertext length with AND gates based on multi-value attributes. Guo et al. [19] discussed a CP-ABE scheme based on AND gates that maintains constant-size decryption keys and is suitable for lightweight devices. Yu et al. [13] first used a scheme with a combination of both CP-ABE and proxy re-encryption (PRE). In this scheme, the authority has re-encryption rights and the ability to update the secret key.

These schemes generate constant-size decryption keys due to their AND-gate structure, but the structure is not expressive enough to hold all conditions.

5.2.3 CP-ABE scheme with LSSS

LSSS is represented as a form of matrix for the attributes in the system. The following CP-ABE schemes were proposed using LSSS gate access structure.

Waters [3] proposed a new methodology of expressing access control with the help of LSSS matrix over the attributes. In this scheme, both encryption and decryption costs are linearly increased because of the access structure's high complexity. It is similar in performance to the scheme proposed by Bethencourt et al. [5]. The algorithm proposed by Waters [3] unnecessarily creates more rows in a matrix from the Boolean formula. Liu and Cao [15] proposed a new procedure to decrease the total number of rows in the LSSS matrix by including Boolean formula as a labeled function to ciphertext. This drastically reduces the costs of communication. Another new scheme designed by Hu et al. [14] improves the efficiency of decentralized ABE scheme with the help of a smaller LSSS matrix. In this matrix, they used better pairing and multi-pairing operations. Naruse et al. [16] described another CP-ABE scheme to address attribute revocation in LSSS access structure using re-encryption done by the cloud. In this scheme, they generate the secret key components for granting attributes instead of generating a new secret key for all attributes.

All these LSSS access structure schemes are in the form of matrices, which increases the size of the ciphertext.

All of the schemes discussed [3–16, 20, 21] are lacking in efficiency due to their access structure representation and are not suitable for IoT devices. Table 5.1 presents a detailed comparison of different ABE schemes.

Table 5.1 Different ABE schemes with different access structures

Scheme	Access structure	Security	Privacy	Access control	Revocation
Bethencourt et al. [5]	Access tree	Selective security	✓	✓	×
Goyal et al. [6]	Bounded access tree	Selective security	✓	✓	×
Ibraimi et al. [9]	Access tree	Selective security	✓	✓	×
Alrawais et al. [1]	Access tree	Selective security	✓	✓	×
Hur and Noh [8]	Access tree	Selective security	✓	✓	✓
Cheung and Newport [10]	AND gates	Selective security	✓	✓	×
Guo et al. [19]	AND gates	Selective security	✓	✓	×
Nishide et al. [12]	AND gates	Selective security	✓	✓	×
Emura et al. [11]	AND gates	Selective security	✓	✓	×
Lewko et al. [20]	LSSS	Full security	✓	✓	×
Waters[3]	LSSS	Selective security	✓	✓	×
Lewko and Waters [21]	LSSS	Selective security	✓	✓	×
Naruse et al. [16]	LSSS	Selective security	✓	✓	✓
Ours(LCP-ABE)	Walsh–Hadamard Transform	Selective security	✓	✓	✓

5.3 Preliminaries

5.3.1 Bilinear maps

Let G_1 and G_T be two multiplicative cyclic groups of prime order p. Let g be a generator of G_1 and e is bi-linear map $e: G_1 \times G_1 \rightarrow G_T$, which has the following properties:

Bilinearity: for all $g, g \in G_I$ and $a, b \in Z_p$, $e(g^a, g^b) = e(g, g)^{ab}$.
Non-degeneracy: $e(g, g) \neq 1$.

5.3.2 CP-ABE framework

Setup: This setup algorithm can take the universal attributes as one parameter, and executed by the key authority it will generate public parameters and a master key.
Encrypt: This encryption algorithm takes the message M and encrypts it by using public parameters and sends it to the cloud via the access structure.
KeyGen: This KeyGen algorithm is executed by the key authority and generates the user's secret key from the user's attribute set.
Decrypt: This algorithm is executed by the user by using the user's secret key from the key authority and the ciphertext received from the cloud.

5.3.3 Access structure

Definition: Let $\{P_1, P_2, \dots P_n\}$ be the set of parties. A collection $A \subseteq 2^{\{P_1, P_2, \dots P_n\}}$ $A \subseteq 2^{\{P_1, P_2, \dots P_n\}}$ is monotone if $\forall Q, R$: if $Q \in A$ and $Q \subseteq C$ then $C \in A$. A monotic access structure is a collection of A of non-empty subsets of $\{P_1, P_2, \dots P_n\}$, that is $A \subseteq 2^{\{P_1, \dots P\}} \backslash \{\varphi\}$. The sets in A are called authorized sets; otherwise unauthorized sets.

In our scheme, we use the Fast Walsh–Hadamard Transform (FWHT) [22, 24] to compute the access structure from the access policy. The FWHT is a divide-and-conquer algorithm that recursively breaks down a truth table of size n into a smaller size, $n/2$. This algorithm takes only $O(n \log n)$ additions or subtractions. We present the construction of the access structure in Section 5.4.3.

5.4 System model

5.4.1 System architecture

Fig. 5.1 represents the system architecture of our scheme. We assume that the communication links from the trusted authority are secure. In our scheme, the following blocks are useful: IoT devices, cloud, -trusted authority, and users. IoT devices are physical objects with internet connectivity that can communicate with other devices over the Internet. These devices perform encryption operations to obtain the ciphertext and transfer it to the cloud for storage purposes. The cloud stores the ciphertext and gives access to users. A trusted authority (TA) generates the keys for encryption and decryption. Then the user gets the ciphertext from the cloud and decrypts it locally.

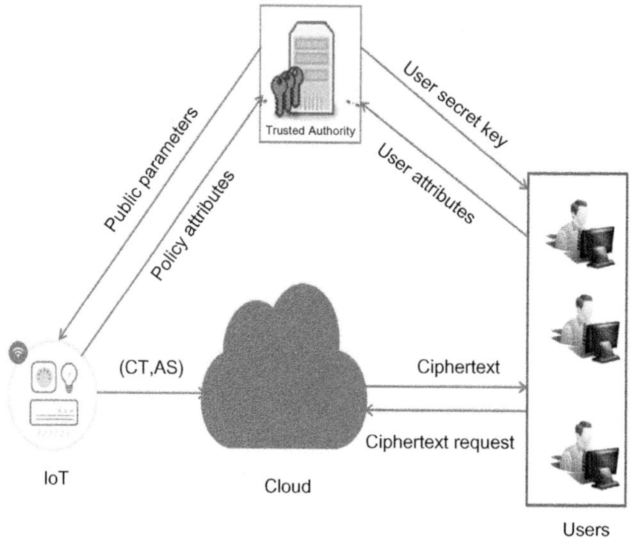

FIG. 5.1

System architecture.

5.4.2 Design goals

We designed a lightweight CP-ABE scheme to achieve the following goals:

- *Privacy*: The data should not disclose to unauthorized persons from the cloud.
- *Access control*: The users can access data if their attributes satisfy the access policy.
- *User revocation*: If any user is removed from the organization, that revoked user should not be able to get the information from the cloud.
- *Lightweight*: The LCP-ABE scheme takes less time for encryption and decryption operations.

5.4.3 LCP-ABE algorithm definitions

The LCP-ABE scheme contains seven algorithms:

- *Setup* (N → PK, MSK): It takes input N as a list of universal attributes and returns a public key PK and master secret key MSK.

- *KeyGen* (MK, $A \rightarrow$ USK): It takes the input as the current MSK, user Attributes and outputs the User secret key USK.
- *Encrypt* (PK, M, WHT $\rightarrow$ CT): It takes input as the PK, message M, access policy represented in the form of access structure WHT and the generates the ciphertext as the CT.
- *ReKeyGen* (δ, MSK $\rightarrow$ rk, PK$'$, MSK$'$): It takes attribute list δ contains the list of the attributes used for update andMSK as input. It returns the new public key PK$'$, new master secret key MSK$'$, and rekey rk.
- *ReEncrypt* (CT, rk, $\delta \rightarrow$ CT$'$): It takes input as the CT, revocable attribute list δ, and the revocable keys rk. It outputs the new ciphertext CT$'$ according to the attributes in δ and by using the revocable keys rk.
- *KeyUpdate* (USK, rk, $\delta \rightarrow$ USK$'$): It takes the input as USK and revocable attribute list δ and outputs the new modified user secret key USK$'$.
- *Decrypt* (CT, PK, USK $\rightarrow M$): It outputs plain text message M based on CT, PK, and the USK as input.

Fig. 5.2 shows the flow of the LCP-ABE algorithms.

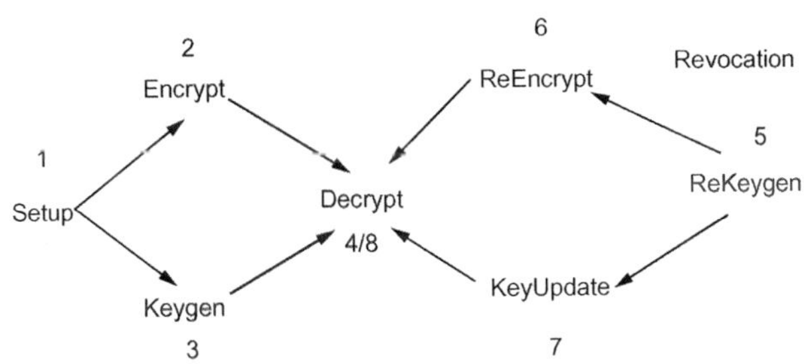

FIG. 5.2

Flowchart of LCP-ABE.

The order of execution of flowchart LCP-ABE algorithms is as follows:

1. TA runs a setup to generate PK and MSK, then sends PK to IoT devices
2. IoT device runs an encrypt algorithm and sends ciphertext along access policy to cloud for storage
3. TA produces a private key for user attributes by running KeyGen algorithm
4. The user runs a decrypt algorithm to decrypt the data based on private key
5. Whenever a user is revoked from system, TA runs ReKeyGen algorithm to generate rekey for re-encryption of ciphertext
6. Proxy server re-encrypts the ciphertext using rekey and sends to cloud
7. Simultaneously, the non-revoked user updates their private keys
8. The user runs a decrypt algorithm to get the plain text using updated secret key

5.4.4 Security game

To prove security strength of LCP-ABE, we define a security game between the Adversary ***Adv*** and a challenger ***Sim*** as follows:

- *Initialization*: The ***Adv*** randomly selects an access structure WHT and transfers it to the ***Sim***.
- Setup: The ***Sim*** executes **setup** phase to generate the public parameters PK and sends to the ***Adv***.
- *Phase 1*: The ***Adv*** sends a query to the **KeyGen** phase by using the user attribute set **A** to get the user secret key USK. But the restriction to this user attribute set is **A** $\Vdash\!\!\!\!/_{\text{WHT}}$ must hold. Then ***Sim*** sends the USK to ***Adv***.
- *Challenge*: The ***Adv*** sends same length of messages M_0 and M_1 to the ***Sim***. Then ***Sim*** randomly selects $\gamma \in \{0, 1\}$ and encrypts M_γ using $_{\text{WHT}}$ to get the ciphertext CT. Finally, *Sim* sends CT along with access structure WHT to the ***Adv***.
- *Phase2*: ***Adv*** repeats actions similar to *Phase 1*.
- *Guess*: The ***Adv*** guesses the value of γ as γ'. The advantage of ***Adv*** is:

$$Adv_{\text{CP-ABE}}^{\text{CPA}}(A) = \left|\Pr[\gamma = \gamma'] - \tfrac{1}{2}\right|$$

5.5 Construction of LCP-ABE

In this section, we construct the proposed LCAP-ABE scheme based on defined algorithms in Section 5.4.3 as follows:

5.5.1 Setup (N ⇢ PK, MSK)

In this algorithm, TA generates PK and MSK as follows:

i. Select a generator g of group G_1 with bilinear map $e: G_1 \times G_2 \rightarrow G_T$
ii. Randomly select $y, t_1, t_2, \ldots t_n \in Z_p$
iii. Compute $Y = e(g,g)^y$, $T_i = g^{t_i}(i \in N)$ for every attribute $i \in N$
iv. Public key $\text{PK} = (e, g, Y, T_i)$ and Master secret key$\text{MSK} = (y, t_i)$

5.5.2 KeyGen (MK, *A* ⇢ USK)

This algorithm perform the following steps to generate the USK from the attribute set $\text{A} = \{a_1, \ldots\ a_n\}$

i. Generate a random number $r \in \mathbb{Z}_p$
ii. Generate bit string (B) from the following formula:
   for i∈ I

if $\text{i} \in \text{A} \wedge \text{i} = \text{i}$ then $b_i = 1$

else $b_i = 0$ then

Bit String $\text{B} = \{b_1, b_1, \ldots\ b_n\}$

For example attributes derived from the policy are ['hcu','idrbt','mtech'] and user attributes are ['hcu','mtech'] then the bit string(B) after computation will be [1,0,1].

iii. Compute $D_1 = g^{y-r}$, $D_w = g^{rt_i}$ where $i \in I$

iv. The USK $= (B, D_1, D_w)$

5.5.3 Encryption (PK, M, WHT → CT)

This algorithm executes in two phases. The first phase constructs the access structure for the given access policy and the second phase generates the ciphertext.

5.5.3.1 Access structure (policy→ WHT)

This algorithm takes input as access policy and converts it into the WHT. The following steps are used in the construction of the access structure.

i. Convert the access policy into the Boolean function.
for example: ((hcu and idrbt) or (hcu and mtech) or (mtech and idrbt)) = ((x1 ∧ x2) ∨ (x1 ∧ x3) ∨ (x3 ∧ x2)).

ii. Extract the attributes I from the access policy I ⊂ N and construct the truth table for the given Boolean function.
for example: From this Boolean function f(x) = ((x1 ∧ x2) ∨ (x1 ∧ x3) ∨ (x3 ∧ x2)) converts to the truth table (TT).

x1	**0**	**0**	**0**	**0**	**1**	**1**	**1**	**1**
x2	0	0	1	1	0	0	1	1
x3	0	1	0	1	0	1	0	1
f(x)	0	0	0	1	0	1	1	1

This table contains the all possible cases of attributes in access policy.

iii. From the TT, we construct the access structure as described in Algorithm 1.

Algorithm 1 Fast Walsh-Hadamard transform

```
Procedure: FWHT(TT )
n ←TT:SIZE()
WHT[n]=TT
for i : 1 → (n/2-1) do
    for j : 0 → n i : do
        WHT[j] := W HT [j] + WHT [j + i]
        WHT [j] := WHT [j] - W HT [j + i]
return (W HT )
```

iv. The pictorial representation of this conversion is shown in the below figure.

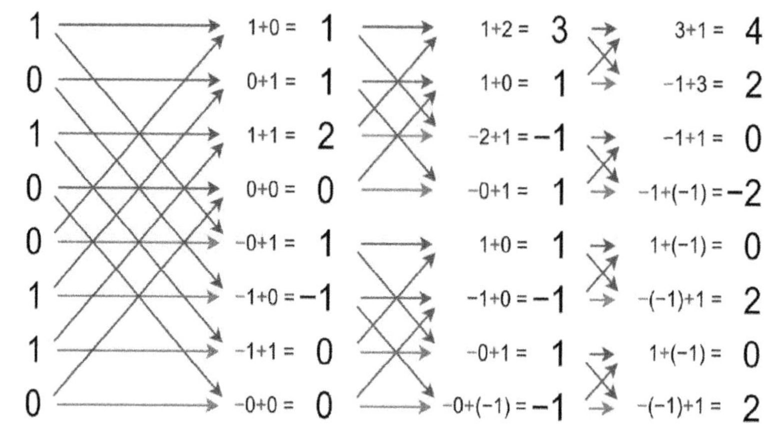

5.5.3.2 Encrypt (PK, M, WHT→ CT)

After construction of the access structure, the encryption algorithm performs the following steps to generate the ciphertext:

i. Randomly select $s \in Z_p$.
ii. Compute $C_1 = M.\, Y^s$, $C_2 = g^s$, $C_w = 1/\left(\prod_{i \in I} T_i\right)$ and the ciphertext element $C_3 = (C_W)^s$.
iii. The ciphertext CT = (WHT, C_1, C_2, C_w, C_3) send to cloud.

5.5.4 ReKeyGen(δ, MSK ⇢ rk)

This ReKeyGen algorithm generate the rk as follows:

i. For each attribute in $i \in \lambda$ randomly choose $t'_1 \in Z_p$
ii. Compute $\mathrm{rk}_i = {t'_i}/{t_i}$ otherwise $\mathrm{rk}_i = 1$
iii. Return the proxy re-key $\mathrm{rk} = \mathrm{rk}_i (i \in N)$ and also
iv. Return new PK' and MSK'

5.5.5 ReEncrypt (CT, rk, δ ⇢ CT′)

This algorithm modifies the ciphertext related to the revoked user to generate new ciphertext CT' as follows:

i. Compute the new attribute related ciphertext component $C'_w = \left(\prod_{i \in I} T_i^{rt_i}\right)^s$ and $C'_3 = (C'_w)^s$
ii. Output the new ciphertext CT′ = (WHT, C_1, C_2, C'_w, C'_3)

5.5.6 KeyUpdate (USK, rk, $\delta \dashrightarrow$ USK′)

The non-revoked users to update their User secret key USK’ as follows:

for each $i \in I$

$D_w = D_w + (t_i \times \text{rk}_i)$

Then $D_2 = g^{(r/D_w)}$

The remaining secret key components are kept the same and output the new USK

$\text{USK'} = (B, D_1, D'_2)$

5.5.7 Decrypt (CT, PK, USK $\dashrightarrow$ M)

In this algorithm, the user decrypts the ciphertext using the secret key after getting ciphertext from the cloud as follows:

i. Compute row of WHT by using the following formula:

$H_m[B]$:B^{th} row of the Hadamard matrix

for i := 0 to WHT. size() - 1

$H_m[B] = (B,i)$

where $B = (p_1, p_2, p_3, \ldots, p_n)$ and $i = (q_1, q_2, q_3, \ldots, q_n)$ are n dimensional binary vectors and $(B,i) = p_1 q_1 + p_2 q_2 + \ldots + p_n q_n$ are i the inner product of vectors B and i.

ii. Compute the acceptance of user attributes by using value of ACC:

$$\text{ACC} := \sum_{j=0}^{\text{WHT.size()}} (\text{WHT}[j] \times H_m[B][j] / \text{WHT.size}()$$

iii. if ACC = 1 then it means that user satisfies the access policy and decryption of the message M as follows:

$$Y_s = e(C_2, D_1).e\left(C_3, \prod\nolimits_{i \in I} D_i\right) Y_s := e(C_2, D_1).e(C_3, \prod\nolimits_{i \in I} D_i)$$
$$M = C_1/Y_s)$$

iv. Otherwise, decryption of the message M is not possible because the User attributes do not satisfy the access policy.

5.6 Security analysis

In this section, we prove the correctness, security strength, and forward secrecy of the LCP-ABE.

5.6.1 Correctness

The correctness of the decryption algorithm is proved as:

$$\text{M} = \frac{C_1}{\ell(C_2, D_1) \times \ell(C_w, D_2)}.$$

$$= \frac{M.Y^s}{\ell(g^s, g^{y-r}) \times \ell\left(\left(\prod_{i\in I} T_i\right)^s, g^{\frac{r}{\sum_{i\in I} t_i}}\right)}$$

$$= \frac{M.\ell(g,g)^{ys}}{\ell(g,g)^{ys-rs} \times \ell\left(g^{\sum_{i\in I} t_i.s}, g^{\frac{r}{\sum_{i\in I} t_i}}\right)}$$

$$= \frac{M.\ell(g,g)^{ys}}{\ell(g,g)^{ys-rs} \times \ell(g,g)^{\sum_{i\in I} t_i.s \cdot \frac{rs}{\sum_{i\in I} t_i}}}$$

$$= \frac{M.\ell(g,g)^{ys}}{\ell(g,g)^{ys-rs} \times \ell(g,g)^{rs}}$$

$$= \frac{M.\ell(g,g)^{ys}}{\ell(g,g)^{ys-rs+rs}}$$

$$= \frac{M.\ell(g,g)^{ys}}{\ell(g,g)^{ys}}$$

$$= \mathrm{M}$$

5.6.2 Security strength

Here, we analyze the security strength of our scheme against two attacks: collusion attack and chosen plain text attack.

5.6.2.1 Collusion attack

In our LCP-ABE scheme, we construct the access structure for each encrypted data and stores in the cloud. The decryption process requires the subset of attributes described by the user. USK generation involves the random number r in the key-generation algorithm for every user. Therefore illegal users can't get the data because the random number is different for every user.

5.6.2.2 CPA security proof

Here, we prove the security strength against CPA through DBDH assumption

Theorem 1 *Suppose our scheme holds the DBDH assumption that no adversary can break the LCP-ABE.*

Proof: This theorem can be proved based on the security game defined in Section 5.4.4. as follows:

Initialization: ***Adv*** chooses the WHT in 2^n form and transfers to the ***Sim***.

Setup: ***Sim*** executes setup phase to provide the PK to the ***Adv***. It takes the input as universal attributes $N = \{A_1, A_2, A_3, \ldots\ldots A_n\}$. ***Sim*** randomly picks the values for each attribute $t_1, t_2, t_3, \ldots t_n$ and generates the PK as $g^{t_1}, g^{t_2}, g^{t_3}, \ldots, g^{t_n}\, g^{t_1}$and outputs PK and MK as $<\ell, g, T, Y>$ and $<y, t>$.

Phase 1: ***Adv*** sends the user attributes A in USK-generation phase where A $\nVdash$ WHT. So, S must be chosen that A can't satisfy the valid combination in WHT. There should be an attribute $i \in I$ that satisfies either $i \in A$ or $i \notin A$. Then, ***Sim*** chooses attribute that satisfies $i \in I$ and $i \notin S$.

Sim chooses the random value $r \in Z_p$.

Bit string (B) is obtained by following formula:

For each attribute $i \in I$:

if $i \in A\ i \wedge i = I$ *then* $b_i := 1$ *else* $b_i := 0$.

Then $B = [b_1, b_2, b_3, \ldots b_n]$.

Then compute secret key values as $D_{1:=}g^{y-r}$, $D_{2:=}g^{\overline{\sum_{i \in I} t_i}^{\,r}}$ and send to ***Adv***.

Challenge: ***Adv*** sends messages with same length M_0 and M_1 $M_0$and$M_1$to ***Sim***. ***Sim*** randomly picks one element from $\mu \in \{0, 1\}$ $\{0,1\}$ and selects the $C_1 = M_\mu$. Z and generates the ciphertext CT*:= (WHT, C_1,C_2,C_w) CT^* WHT, C_1, C_2, C_w send to ***Adv***.

Phase 2: ***Adv*** repeats *Phase 1*.

Guess: ***Adv*** guess a output γ' of γIf $\gamma' = \gamma$ ***Sim*** generates the output 0 to indicate that $Z = g(g, g)^{abc} g^{abc}$ is a valid ciphertext, so the advantage of Adv is ε. Thus

$$P_r[\text{Sim} \rightarrow \text{“0”} \mid Z = e(g,g)^{abc}] = P_r[\gamma' = \gamma \mid Z = e(g,g)^{abc}] = \tfrac{1}{2} + \varepsilon$$

If $\gamma' \neq \gamma$, then sim outputs 1 to indicate $Z = e(g,g)^z$. Then it's completely random from ***Adv***. Thus.

$P_r[\text{Sim} \rightarrow \text{“1”} | Z = e(g,g)^z] = P_r[\gamma' - \gamma \mid Z - e(g,g)^z] = \tfrac{1}{2}$

Based on the preceiding analysis, the advantage of ***Adv*** to break our scheme is:

$$\tfrac{1}{2}\left(\tfrac{1}{2} + \varepsilon + \tfrac{1}{2}\right) - \tfrac{1}{2} = \tfrac{\varepsilon}{2}.$$

Hence, any polynomial time adversary cannot break the LCP-ABE scheme.

5.6.3 Forward Secrecy

Our scheme guarantee the forward secrecy of the IoT data against any users contraining the revoked users. Forward secrecy means that session keys will not be compromised even if the private key of the server is compromised. Revoked users cannot access the data in the future after the revocation happens. In our scheme, after the revocation of the particular user, the trusted authority recomputes the master key, public parameters, and RK. The proxy updates ciphertext components and generates the new ciphertext. Thus, revoked user who can have been revoked, cannot able to find the blinding factor.

$$e(C_2, D_1) \times e\ (C_w, D_2)$$

So, the forward secrecy of the IoT data against revocation is assured.

5.7 Performance analysis

This section measures the communication and computation overhead theoretically.

5.7.1 Communication overhead

Here, we mainly focus on size of ciphertext. The secret keys are generated after the execution of encryption and KeyGen algorithms, respectively. Table 5.2 presents the comparison of LCP-ABE with the existing schemes with parameters like ciphertext size and secret key size. In Table 5.2, N is the list of the universal attributes, ϕ represents the list of attributes in the access policy, w represents the user attributes, and S_{G_0}, S_{G_1}, and S_{G_T} represents the size of each element in the Groups G_0, G_1, and G_T respectively. B represents the attributes in the access policy.

As shown in Table 5.2, our scheme generates three group elements in ciphertext and two group elements along with the bit string in USK, which is less than existing schemes. Our scheme takes minimal time to communicate the messages. Thus our scheme is more efficient than existing schemes [8, 13, 16]

Table 5.2 Communication overhead

Scheme	Ciphertext size	Secret key size
Hur and Noh [8]	$(2\phi+1).S_{G_0}+S_{G_1}$	$(2w+1).S_{G_0}$
Yu et al. [13]	$(N+1).S_{G_0}+S_{G_T}$	$(2w+1).S_{G_0}$
Naruse et al. [16]	$S_{G_0}+(2\phi+1).S_{G_{T0}}$	$w.S_{G_0}$
LCP-ABE	$2.S_{G_0}+S_{G_T}$	$\|B\|+2.S_{G_1}$

5.7.2 Computation overhead

Computation overhead depends on expensive operations like exponentiation on the group element and the pairing computations involved in the execution of the encryption, decryption, and keyGen algorithms. Table 5.3 presents the computation costs of the LCP-ABE scheme and existing schemes [8, 13, 16].

Table 5.3 Computation overhead

Scheme	Encrypt	KeyGen	Decrypt
Hur and Noh [8]	$(2\phi + 1).E_G + P_e$	$(2w+ 1).E_G$	$(2w + 1).P_e + \phi_\gamma.E_G$
Yu et al. [13]	$(N + 2).E_G$	$(2w + 1).E_G$	$(N + 1).P_e$
Naruse et al. [16]	$(2\phi + 1).P_e + 2\Phi. E_G$	$w.E_G$	$2w.P_e + E_G$
LCP-ABE	$(\|WHT\| + 2).E_G$	$2.E_G$	$2.P_e$

In Table 5.3, N represents the universal attributes. E_G represents the number of exponentiation operations performed on the elements of the group, P_e is the number of pairing operations, and ψ represents the number of non-leaf nodes in the access tree.

As shown in Table 5.3, our scheme takes WHT + 2, EG number of operations for the encryption algorithm and two exponentiation operations and two pairing computations required for the KeyGen decryption algorithm, respectively. As such, our scheme takes minimal time to perform all the algorithms.

5.7.3 Experimental results

We performed simulation experiments on a laptop with an Intel i5 2.3 GHz CPU with 8 GB RAM and the Ubuntu 16.04 operating system. We also used an Android 6.0.1 smartphone with quad core processor and 2 GB memory as the IoT device. To implement schemes, we used Python programming language by using Charm-crypto 0.43 Library [23]. Here, we measure the computational cost of our scheme with the help of Figs. 5.3–5.5.

Fig. 5.3 shows the time required to perform the encryption based on different attribute sets. From Fig. 5.3, we can observe that time for encryption increases linearly along with the number of attributes in the access policy.

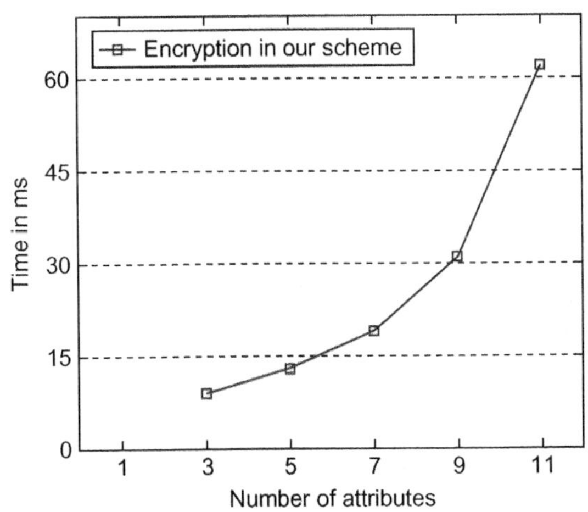

FIG. 5.3

Encryption running time.

Fig. 5.4 shows the time required to perform the KeyGen operation w.r.t to the attributes given by the user. Time for the KeyGen operation changes linearly up to some level and then requires only constant time to generate the USK.

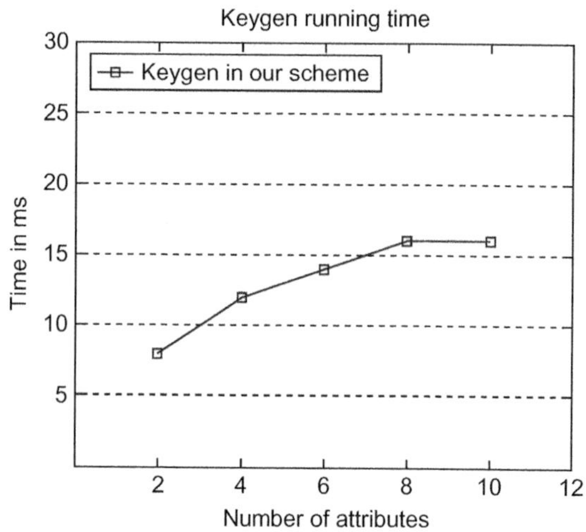

FIG. 5.4

Key generation running time.

Fig. 5.5 shows the time required to perform the decryption operation. Our scheme takes the same amount of time to perform decryption regardless of the number of attributes. It also computes the message in constant time because of the access structure.

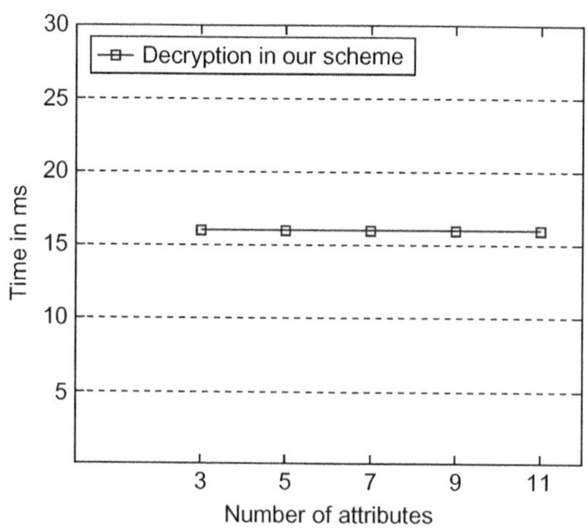

FIG. 5.5

Decryption running time.

5.8 Conclusion

In this chapter, we proposed a *Lightweight CP-ABE* scheme for data privacy and access control in cloud-assisted IoT. We used a new access structure called the Walsh–Hadamard transform that supports user revocation. Our scheme reduced the computation overhead of IoT devices during encryption and allows the occurrence of a single attribute multiple times in access policy during decryption. LCP-ABE proved that it is resistant to collusion attack and chosen plain text attack and maintained forward secrecy. Through performance analysis, we proved that our scheme is lightweight and more suitable for IoT devices. In further work, we extend LCP-ABE to support authentication of IoT devices, integrity of data, and dynamic data operations for IoT data in the cloud.

References

[1] A. Alrawais, A. Alhothaily, C. Hu, X. Cheng, Fog computing for the internet of things: Security and privacy issues. IEEE Inter. Comput. 21 (2) (2017) 34–42, https://doi.org/10.1109/MIC.2017.37.

[2] J. Singh, J. Bacon, H. Ko, T. Pasquier, D. Eyers, Twenty security considerations for cloud-supported IoT. IEEE Internet Things J. 3 (3) (2016) 269–284, https://doi.org/10.1109/JIOT.2015.2460333.

[3] B. Waters, Ciphertext-policy attribute-based encryption: An expressive, efficient, and provably secure realization, in: Proceedings of Public Key Cryptography, Vol. 6571, 2011, pp. 53–70.

[4] A. Alrawais, A. Alhothaily, C. Hu, X. Cheng, An attribute-based encryption scheme to secure fog communications. IEEE Access 5 (2017) 9131–9138, https://doi.org/10.1109/ACCESS.2017.2705076.

[5] Bethencourt, J., Sahai, A., and Waters. B.(2007). Ciphertext-policy attribute-based encryption. In Proceedings of the 2007 IEEE Symposium on Security and Privacy, Vol. 24 of SP'07, pages 321–334. IEEE Computer Society, 2007. ISBN 0-7695-2848-1. doi: https://doi.org/10.1109/SP.2007.11.

[6] V. Goyal, R.A. Jain, O. Pandey, A. Sahai, Bounded ciphertext policy attribute based encryption. in: I. Damgård, L.A. Goldberg, M.M. Halldorsson, A. Ingolfsdottir, I. Walukiewicz (Eds.), Automata, Languages and Programming, Springer, Berlin/Heidelberg, 2008, ISBN: 978-3-540-70583-3, pp. 579–591, https://doi.org/10.1007/978-3-540-70583-3n.

[7] Z. Guan, J. Li, L. Wu, Y. Zhang, J. Wu, X. Du, Achieving efficient and secure data acquisition for cloud-supported internet of things in smart grid, IEEE Internet Things J. 4 (6) (2017) 1934–1944, https://doi.org/10.1109/JIOT.2017.2690522.

[8] J. Hur, D.K. Noh, Attribute-based access control with efficient revocation in data outsourcing systems, IEEE Trans. Parallel Distrib. Syst. 22 (7) (2011) 1214–1221, https://doi.org/10.1109/TPDS.2010.203.

[9] Ibraimi, L., Tang, Q., Hartel, P. and Jonker, W.(2009). Efficient and provable secure ciphertext-policy attribute-based encryption schemes. In Proceedings of the 5th International Conference on Information Security Practice and Experience, ISPEC'09, pages 1–12. Springer-Verlag. ISBN 978-3-642-00842-9. doi: https:/doi.org/10.1007/978-3-642-00843-6n.1.

[10] Cheung, L., and Newport, C. Provably secure ciphertext policy abe. In Proceedings of the 14th ACM Conference on Computer and Communications Security, CCS '07, pages 456–465. ACM, 2007. ISBN 978-1-59593-703-2. Doi:https:/doi.org/10.1145/1315245.1315302.

[11] K. Emura, A. Miyaji, K.O. Nomura, M. Soshi, A ciphertext-policy attribute-based encryption scheme with constant ciphertext length, in: F. Bao, H. Li, G. Wang (Eds.), Information Security Practice and Experience, Springer, Berlin/Heidelberg, 2009, ISBN: 978-3-642-00843-6, pp. 13–23.

[12] T. Nishide, K. Yoneyama, K. Ohta, Attribute-based encryption with partially hidden encryptor-specified access structures, in: S.M. Bellovin, R. Gennaro, A. Keromytis, M. Yung (Eds.), Applied Cryptography and Network Security, Springer, Berlin/Heidelberg, 2008, ISBN: 978-3-540-68914-0, pp. 111–129.

[13] Yu, S., Wang, C., Ren, K., and Lou, W (2010). Attribute based data sharing with attribute revocation. In Proceedings of the 5th ACM Symposium on Information, Computer and Communications Security, ASIACCS '10, pages 261–270. ACM, ISBN 978-1-60558-936-7. doi: https:/doi.org/10.1145/1755688.1755720.

[14] Hu, C., Zhang, F., Xiang, T., Li, H., Xiao, X. and Huang, G.(2014). A practically optimized implementation of attribute based cryptosystems. In 2014 IEEE 13th International Conference on Trust, Security and Privacy in Computing and Communications, Pages 197–204, doi: https:/doi.org/10.1109/TrustCom.2014.29 (web archive link).

[15] Liu, Z., and Cao, Z. (2010), On efficiently transferring the linear secret-sharing scheme matrix in ciphertext-policy attribute-based encryption. 2010:374

[16] T. Naruse, M. Mohri, Y. Shiraishi, Attribute-based encryption with attribute revocation and Grant function using proxy re-encryption and attribute key for updating, in: J. Park, I. Stojmenovic, M. Choi, F. Xhafa (Eds.), Future Information Technology. Lecture Notes in Electrical Engineering, Vol. 276, Springer, Berlin/Heidelberg, 2014.

[17] P. Kumar, S. Kumar, A.P.J. Alphonse, Attribute based encryption in cloud computing. J. Netw. Comput. Appl. 1084-8045108 (C) (2018) 37–52, https://doi.org/10.1016/j.jnca.2018.02.009.

[18] A. Shamir, How to share a secret, Commun. ACM 0001-078222 (11) (1979) 612–613, https://doi.org/10.1145/359168.359176.

[19] F. Guo, Y. Mu, W. Susilo, D.S. Wong, V. Varadharajan, Cp-abe with constant-size keys for lightweight devices, IEEE Trans. Inf. Forensics Security 1556-60139 (5) (2014) 763–771, https://doi.org/10.1109/TIFS.2014.2309858.

[20] Lewko, A. B. Okamoto, T., Sahai, A., Takashima, K., and Waters, B. Fully secure functional encryption: Attribute-based encryption and (hierarchical) inner product encryption", Proceedings of EUROCRYPT, 2010, Vol. 6110, pp. 62–91.

[21] A. Lewko, B. Waters, New proof methods for attribute based encryption: Achieving full security through selective techniques, in: 32nd International Conference on Cryptology (CRYPTO 2012), Springer, Santa Barbara, 2012, , pp. 180–198.

[22] C.-K. Wu, D. Feng, Boolean Functions and Their Walsh Transforms, Springer, Berlin/Heidelberg, 2016, ISBN: 978-3-662-48865-2, pp. 1–30, https://doi.org/10.1007/978-3-662-48865-2n.1.

[23] A. Akinyele, C. Garman, I. Miers, M.W. Pagano, M. Rushanan, Green, and A. D. Rubin. Charm: a framework for rapidly prototyping cryptosystems, J. Cryptog. Eng. 2190-85163 (2) (2013) 111–128, https://doi.org/10.1007/s13389-013-0057-3.

[24] B.J. Fino, V.R. Algazi, Unified matrix treatment of the fast walsh-hadamard transform, IEEE Trans. Comput. 0018-934025 (11) (1976) 1142–1146, https://doi.org/10.1109/TC.1976.1674569.

CHAPTER

Soft sensor with shape descriptors for flame quality prediction based on LSTM regression

6

K. Sujatha*, N.P.G. Bhavani[†], V. Srividhya[†], V. Karthikeyan*, N. Jayachitra[‡]

**Department of EEE, Dr. M.G.R. Educational & Research Institute, Chennai, India* [†]*Department of EEE, Meenakshi College of Engineering, Chennai, India* [‡]*Department of Chemical Engineering, Dr. M.G.R. Educational & Research Institute, Chennai, India*

6.1 Introduction

It is stated that in coal fired plants, burners are significant equipment, center for energy generation systems [11]. Searing air and steam combined together rotate the turbines. Ultimately electrical energy is generated by the generator coupled to the turbines. If the injection of pulverized coal (fuel) into the combustion chamber is left unmonitored the boiler is prone to the risk of explosion. In such a highly crucial environment with boilers whose combustion chambers in thermal and gas turbine power plants have a temperature nearly 1200°C, ordinary gas analyzers or temperature sensors are not suitable devices, hence an optical flame detector system is needed. The flame monitoring system displays the status of the flame at the control room so that the inlet fuel can be cut off so that the flame goes out, lowering the risk of high accumulation of fuel resulting in explosion of the boiler.

The aim of our system [11] is to monitor the furnace flame using infrared cameras. To automate this combustion, quality-estimation artificially intelligent (AI) algorithms are used to determine the features of the furnace flame that correlate with air-fuel ratio, NO_x, CO, and CO_2 emission levels, temperature, and so on. The 3D temperature profiler is designed to provide control of furnace and flame temperature, which also reduces the flue gas emissions. This is key in achieving high combustion quality as indicated by Sujatha et al. [11]. The system is also designed to provide guidance for balancing the air-fuel ratio so as to ensure complete combustion. The main goal is to introduce a continuous monitoring scheme online proposed in [11]. This will ensure a safe and secured combustion for addressing the escalating demands to attain high thermal efficiency at the furnace level. It will also help to

Real-Time Data Analytics for Large Scale Sensor Data. https://doi.org/10.1016/B978-0-12-818014-3.00006-1

reduce flue gas emissions and thereby improve combustion quality. The proposed scheme can be used dynamically to automatically vary the air-fuel ratio based on the color of the flame images (feedforward control). An intelligent feedforward control and flame image analysis were implemented to adjust the air-fuel ratio and to minimize flue gas emissions, thus ensuring complete combustion. The systems have been evaluated in both the laboratory as well as industrial-scale combustion rigs under a variety of operating conditions.

Apart from energy production, the world is looking at ways to reduce global warming. Among various industrial sectors the power plant plays a major role in energy production, but it also causes global warming. Thus it is important to reduce hazardous gases from escaping into the atmosphere. Global warming is a very serious problem leading to depletion of the ozone layer. The deteriorating ozone layer allows the passage of harmful ultraviolet rays to raise the global average temperature near the Earth's surface. The flue gases from thermal and gas turbine power plants contain a mixture of CO, CO_2, NO_x, and SO_x (greenhouse) gases, which contribute to changing climate patterns.

As is extensively established across the globe, generating electricity by burning coal is the root cause of troubles connected to the ecological conditions of operating power plants [11]. The thermal power plant symbolizes the huge contributor to acid rain of any industrial sectors, since they serve as the foundation for emission of sulfur oxides [12].

According to Sujatha et al. [17], coal-fired industrial applications are also a noteworthy cause of nitrogen oxides, with an impact similar to that of vehicle exhaust. The burning of coal sturdily adds to acid rain and greenhouse gases, leading to global warming [16]. Regrettably, beyond its ecological impact, burning coal has an impact on human health. Exposure to toxic elements that cause air pollution can result in cardiac diseases, breathing problems, and lung cancer, among other human health issues. Air pollution from thermal power stations is also connected to other diseases, such as newborn death, chronic bronchitis, asthma, undesirable reproductive effects, and other lung diseases. An elucidation for monitoring and efficiently reducing SO_x and NO_x emissions from power plants when their levels exceed the admissible limits proves to be fruitful when such an automated scheme is implemented. For our proposed automated system, we designed a wireless embedded computing system to control and reduce SO_x a NOx emissions, which, as already discussed, are a major cause of climate change.

This technique was incorporated [19, 20] after preprocessing and extracting the features from flame images, which are initially collected from the combustion chamber of the boiler in the power station. The images are used to estimate ignition quality and SOx and NOx outflows. We used radial basis function (RBF) and Fisher linear discriminant radial basis function (FLDRBF) algorithms. Testing results demonstrate that FLDRBF gives great order execution when compared with other algorithms like error detection and correction (EDC) and Fisher linear discriminant analysis (FLD). Our results [14] are supported by using the usual algorithms like

RBF and backpropagation algorithm (BPA) for combustion quality and SO_x and NO_x emissions monitoring. BPA, which is a type of multilayer perceptron network as proposed in [3].

The power plant turbine is one of the significant core elements in energy generation. Malfunctioning in the turbine system might decrease the efficiency of the plant, which can lead to damage and failure of the system. This, in turn, results in increased costs and creates a major issue with combustion quality, which ultimately leads to pollution in the environment. Therefore turbine monitoring needs to be mandatory in order to avoid damage caused by indecorous combustion in the plant system. As such, we focus on the monitoring and control of the combustion unit of the turbine by using image processing and AI techniques.

We designed a flame monitoring system using AI methods to categorize the flame attributes that are correlated with NO_x emissions, SO_x levels, temperature, and so on. Among the 102 flame images generated from the boiler power control room, 51 were taken for training and 51 one were used for testing purposes. In this section, a simple parallel architecture was used. The classifier FLD analysis did not affect performance and other classifiers were used, such as BPA, RBF, and PRBFBPA. Because we used different feature combinations for categorization, we found misclassifications in the category of combustion output. However, we found that inputting three attributes to BPA and inputting four attributes to RBF showed better classification patterns of combustion quality. The performances were compared for accuracy and recall for the same classes (1, 2, and 3). Apparently, it supports the parallel architecture of the intelligent classifiers for combustion quality. However, the recall was closer to 1, but in the case of precision there was presence of partial or incomplete combustion. The value for both recall and precision were not found to be same, which might be due to high complexity of the architecture. Finally, when the comparison estimation was performed with test and validation results, it showed that partial and incomplete combustion had taken place. Simultaneously, the error value of 0.4 was observed in class 3 in case of comparison between target and actual values for the different categories of flame images. However, we were able to enhance classification performance through preprocessing of the acquired images.

By including entropy for all the classes we found that the values for white balance lesser than the other two stages. Hence, the significant of image processing as showed a value based addition for predicting the complete combustion in the gas turbine. In order to avoid error and allow complete combustion, we needed to define another classifier. We included the feature extraction process for each specific image with selected parameter.

From our observation, the inputs of the images are more significant for the analysis of combustion quality. Hence we attempted to improve monitoring by including feature extraction. Sections of flame images (51 flame images) for a variety of combustion conditions were put together from the centralized area. Cluster 1 refers to complete combustion (flame 1 to flame 18), cluster 2 indicates partial combustion (flame 19 to flame 38), and cluster 3 refers to incomplete combustion (flame 39 to flame 51). The mining of the features was done using 2D and 3D methods based on boundary detection being the most favorable set of attributes for flame image analysis. Continuing with feature extraction, we further classified the output of the edge detection process using the FNN method in order to estimate NO_x and SO_x emissions and combustion quality.

Further, the performance metrics of BPA using 23 attributes showed better output. However, recall value of the flame images showed a slight reduction in precision thus corresponds to the emissions of SO_x and NO_x in the gas turbine. Therefore an intelligent sensor for assessment of SO_x and NO_x emissions using Internet of Things (IoT) and big data analytics proposed by Das et al. (2019) could be made possible from flame image analysis for real time monitoring [1]. As a result, an intellectual system can be designed to check and control the air-fuel ratio. The SO_x and NO_x emissions are also minimized thereby reducing air pollution. However, we found only a 19% reduction in emissions of NO_x and SO_x to the environment. Thus we are in need of specific classifiers that can monitor and control the system such that combustion quality also improves along with the reduction of SO_x and NO_x emissions.

The last few years [5] have seen extraordinary advancements, including various virtualization and system advancements, benefit situated models, and so on, prompting calculation originals, for example, framework figuring [2]. In any case, advancing applications, for example, bio-therapeutic science and the IoT, have quickened novel information serious calculation ideal models. Distributed computing is advertised as the eventual fate of data-innovation ideal models in that it can use processing as a utility by giving huge computational and informational assets in virtualized conditions. Apache Hadoop is a collection of open-source software utilities that facilitate using a network of many computers to solve problems involving massive amounts of data and computation. Hadoop, based on the guide diminish worldview, gives gigantic information the board skill to information serious employments. It licenses clients to misuse calculation and information assets of a Hadoop bunch through its guide diminish execution, dealing with various low-level issues, for example, correspondence, adaptation to non-critical failure, and information the board among others and giving a straightforward information the executives interface. Attributable to execution enhancement and better dependability, Hadoop separates a document into a few information squares and places them in various hubs inside the Hadoop bunch.

However, Hadoop does not monitor arrangement of information hinders among group hubs in the wake of part an information record while settling on choices with respect to planning of occupations. Each activity has its very own information record necessity and in this manner planning employments in a position negligent way prompts runtime information developments among the Hadoop group. Despite what might be expected, if employments could be booked among Hadoop group hubs with from the earlier data of the area of information squares they require, at that point such superfluous information developments could be diminished at runtime. Such a shrewd information position and resulting work planning has been seen to enhance information calculation co-area and has been a demonstrated strategy to enhance execution.

Distributed computing is an emerging field of research [2] and is helpful for all types of clients, from end clients to top business organizations [3]. There are a few areas of research in distributed computing, including load adjusting, cost the board, work process booking, and so forth. To manage such issues, some regular strategies are produced, some of which are not successful. The utilization of nature-motivated improvement in distributed computing is a noteworthy zone of concern. In this section, a point-by-point (yet short) overview covers the pertinence of nature-propelled calculations in different distributed computing issues. Part of the section is devoted to learning about calculations based on biological activities and mentioned issues of distributed computing [4]. We also present some ideas for future research in distributed computing and other applications.

Intrusion-detection systems provide a secured environment for data transfer. The major problem lies in handling complex datasets that are varied in nature [5, 6]. The unequal distribution causes classification that is dependent on the majority class. Hence each layer is modeled to encounter the changes due to intruder attack using particle swarm optimization (PSO), genetic algorithms, and other bio-inspired algorithms.

As a part of the previous research work detection classifier for monitoring and controlling the combustion process using ACO is incorporated. Samples of 51 flame images corresponding to various combustion conditions were collected from the control room. We compared combustion quality and NO_x and SO_x emissions using a feedforward neural network (FFNN) using BPA and ACO. From the performance matrices, ACO showed better results when compared with precision and recall values. However, both BPA and ACO were used with samples from the output of the edge-detection method. ACO performances showed slight variation in precision and recall, thus proving that it is better for monitoring and controlling combustion quality and therefore may subsequently reduce NO_x and SO_x emissions.

6.2 Literature survey

In the previous proposed work [11] studied to modify or alter the process according to our need. The Maximum Posterior Marginal (MPM) method based on the Hidden Markov Model (HMM) was used to identify the combustion conditions. Another method based on edge detection and pattern recognition was used to identify the combustion conditions. Though there are techniques to infer flame temperature, these techniques do not provide data regarding CO_2 and NO_x emissions in flue gases thus offering a control of air/fuel ratio.

Our aim [11] is to devise a flame-based supervisory expert system using a series of cameras and AI methods in order to categorize flame features that can be interrelated with the air-fuel ratio, flue gas emission levels, temperature, and so on.

The 3D temperature models were premeditated to reduce the flue gas emissions by managing the furnace and flame temperature, which is key in achieving high combustion quality. The main goal is to address the exponentially growing demands for high output from thermal power plants by online monitoring and further reducing flue gas emissions and improving combustion quality.

It is confessed that steam is used for driving the turbines in turn coupled to a generator which generates power in the power station [12]. The vessel that contains the pressurized water in a boiler is typically made of steel (alloy steel) or wrought iron. Copper is used for making the furnace because of its better formability and high thermal conductivity; however, in the modern era, the high cost of copper often makes it uneconomical as a substitute.

For example, the literature [7–11] adopts an improved support vector machine (SVM) to increase the effectiveness and accuracy of the prediction as well as to reduce its running time. According to [12] the Gray Modal (GM) (1,1) with HMM reduces the prediction error and allows prediction based with small amount of data. Other literature [13–15] uses a variety of neural network-combined algorithms to improve the accuracy of predictions. Still other literature [16–18] reconstructs the combustion quality as a time series based on chaos theory and fuzzy time series theory to improve prediction accuracy. These methods have greatly improved accuracy and reliability of prediction models. However, these traditional methods only verify the flame shape at the current time and do not predict the combustion quality at the next moment.

In recent years, with the development of big data and AI, deep learning has been applied to many fields with good results. Among them, Long Short-Term Memory (LSTM)) has been used in the power industry [19, 20], text analysis [21, 22], the biomedical field [23], and book management [24]. In essence, LSTM is a special recurrent neural network. The biggest feature of this neural network is that the output of the last iteration is inextricably linked to the output after the next processing. This feature makes it ideal for processing data that is continuous and interacting on the time axis.

In view of the fact that traditional technology does not make a guiding prediction of combustion quality in the next time period, this chapter proposes a LSTM combustion quality prediction model based on flame shape, with the minimum objective function as the optimization target. It is excellent for obtaining an optimal prediction model. The model can predict the combustion quality based on flame shape in the next time period. LSTM is an excellent upgrade model for RNN, with most of the advantages of the RNN model, as compared with gradient back propagation process. LSTM is suitable for dealing with issues that are highly correlated with time series, such as acoustic translation, generation of dialogue, encoding/decoding, and so on. Its structure is shown in Fig. 6.1.

The saturated, superheated water vapor is heated to the required temperature of 5400°C in the primary and secondary super heaters and fed into the High Pressure Turbine (HPT). The hot gas leaving the furnace transfers heat by radiation and convection to the Primary Super Heater (PSH) or first stage of superheat, the Secondary Super Heater (SSH) or second stage of superheat, and the reheaters in succession. Before leaving the chimney, the hot gases are sucked away by the Induced Draft

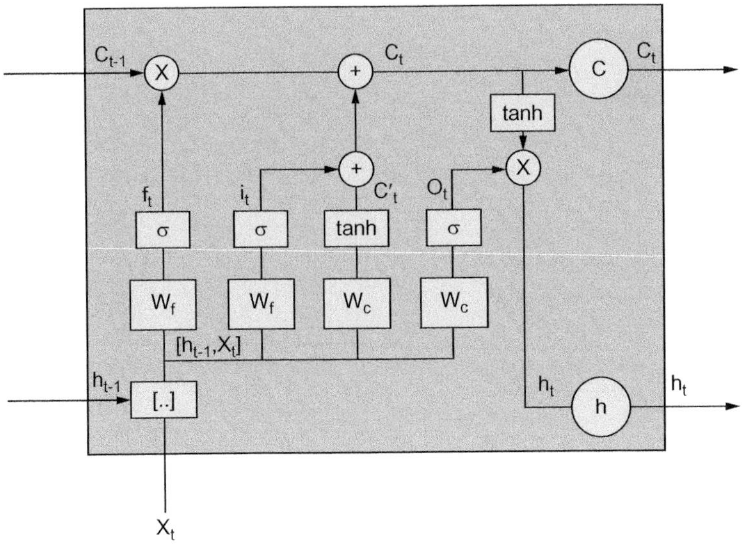

FIG. 6.1

LSTM unit structure diagram.

Adapted from D. Gan, Y. Wang, N. Zhang, W. Zhu, Enhancing short-term probabilistic residential load forecasting with quantile long–short-term memory, J. Eng. 2017(14) (2017) 2622–2627.

(ID) fan, which delivers heat to the feed water in the economizer. The Forced Draft (FD) fan provides the required air for the combustion process.

A desuperheating spray is introduced between the primary (first stage super heater) and secondary super heater (second stage super heater) sections to control steam temperature. Cold water is sprinkled on the exit flue gas so that the gas analyzer placed at the exit point is not affected. The furnace heat absorbed by the super heater or reheater is also controlled by tilting the burners or changing the pattern of firing by selecting the burner. The gas recirculation is used to increase the mass flow of hot gases through the super heaters by withdrawing some amount of gas from the gas outlet into a manifold around the bottom of the furnace chamber. This tends to cool the walls and increase heat absorption by the super heaters.

6.3 Description of flame shape and burner system

The actual flame shape and the typical fuel supply/distribution system in a coal-fired power plant are shown in Fig. 6.2. In thermal power plants, the burners are V94.2 Ansaldo (120 MW). Each combustion chamber has 20 burners, of which 12 are Lignite burners and eight are oil burners. Two flame detectors in each chamber send commands via the monitoring system that detects the flameout condition. High temperature has a drastic effect on main parts of the detector, so it is installed out of this region and isolated by an external lens. Cooling and soot-blowing air are used to avoid overheating and clogging of the line of sight.

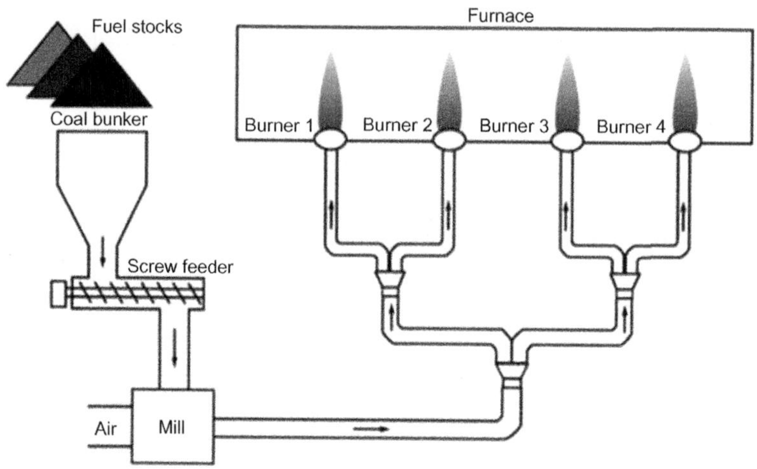

FIG. 6.2

Fuel supply/distribution system.

From Y. Yan, Advanced monitoring and process control technology for coal-fired power plants, in: D. Roddy (Ed.), Advanced Power Plant Materials, Design and Technology, 2010, pp. 264–288.

Valuable information on combustion quality and emissions can be extracted from viewing flame shape and color. For example:

- Flame too long: loose or cracked fuel oil burner nozzle, return line clogged
- Flame too long and peripherally irregular: worn or partially clogged fuel nozzle
- Flame too short and peripherally irregular: natural gas supply lines partially clogged

Various flame types are shown in Fig. 6.3A–D.

An image visualization system could be used and analyzed with processing software and formulas to best predict and improve plant performance.

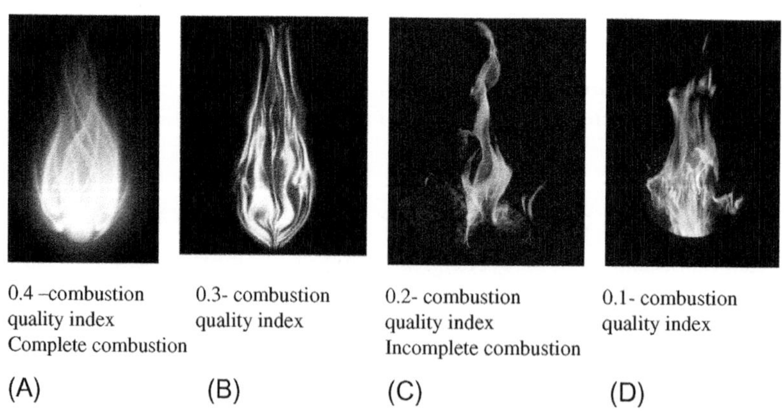

FIG. 6.3

Categorization of flame shapes from a flame detector. (A) Flame normal conditions, (B) flame too long, (C) flame too long and peripherally irregular, (D) flame too short and peripherally irregular.

6.4 LSTM for flame shape-based combustion quality prediction model

6.4.1 Model construction

The model consists of an input layer, hidden layer, and output layer. The input layer is mainly used for preprocessing and dataset division of the original data. The hidden layer is trained based on the training set through the Adam optimizer introduced. The optimization parameters are optimized according to the principle that the loss value is the smallest. The output layer predicts the data according to the model learned in the hidden layer and performs data restoration for the scaling of the previous data preprocessing. Model tuning is also done for obtaining the optimal output. The LSTM flowchart and prediction model framework is shown in Fig. 6.4A and B, respectively.

6.4.2 Model training and prediction

The model training mainly takes the hidden layer as the research object, and the raw data input in the input layer can be expressed as:

$D = [[G_1, G_2, \ldots, G_n],[T_1, T_2, \ldots, T_n], [P_1, P_2, \ldots, P_n], [F_1, F_2, \ldots, F_n]]$, dividing the raw data into training Set $Tr = [[G_1, G_2, \ldots, G_m], [T_1, T_2, \ldots, T_m], [P_1, P_2, \ldots, P_m], [F_1, F_2, \ldots, F_m]]$, test set $Te = [[G_{m+1}, G_{m+2}, \ldots, G_n], [T_{m+1}, T_{m+2}, \ldots, T_n], [P_{m+1}, P_{m+2}, \ldots, P_n], [F_{m+1}, F_{m+2}, \ldots, F_n]]$, look_back $= [[G_{m-a}, \ldots, G_{m-2}, G_{m-1}], [T_{m-a}, \ldots, T_{m-2}, T_{m-1}], [P_{m-a}, \ldots, P_{m-2}, P_{m-1}], [F_{m-a}, \ldots, F_{m-2}, F_{m-1}]]$, $a < m < n$, where f_t, i_t, C_t, and O_t are the forgetting gate, input gate, unit state, and output gate output timing, respectively; 'W', 'b', and 'tan h' are the corresponding weights, deviations, and excitation functions, respectively. The LSTM model training process is given by Eqs. (6.1)–(6.7). The flowchart for the LSTM model is shown in Fig. 6.4A.

1. Information screening and discard:

$$f_t = \delta\left(w_f \bullet [h_{t-1}, x_t] + b_f\right) \tag{6.1}$$

2. Add new information:

$$i_t = \delta(w_i \bullet [h_{t-1}, x_t] + b_i) \tag{6.2}$$

And

$$C' = \tanh\left(w_c \bullet [h_{t-1}, x_t] + b_c\right) \tag{6.3}$$

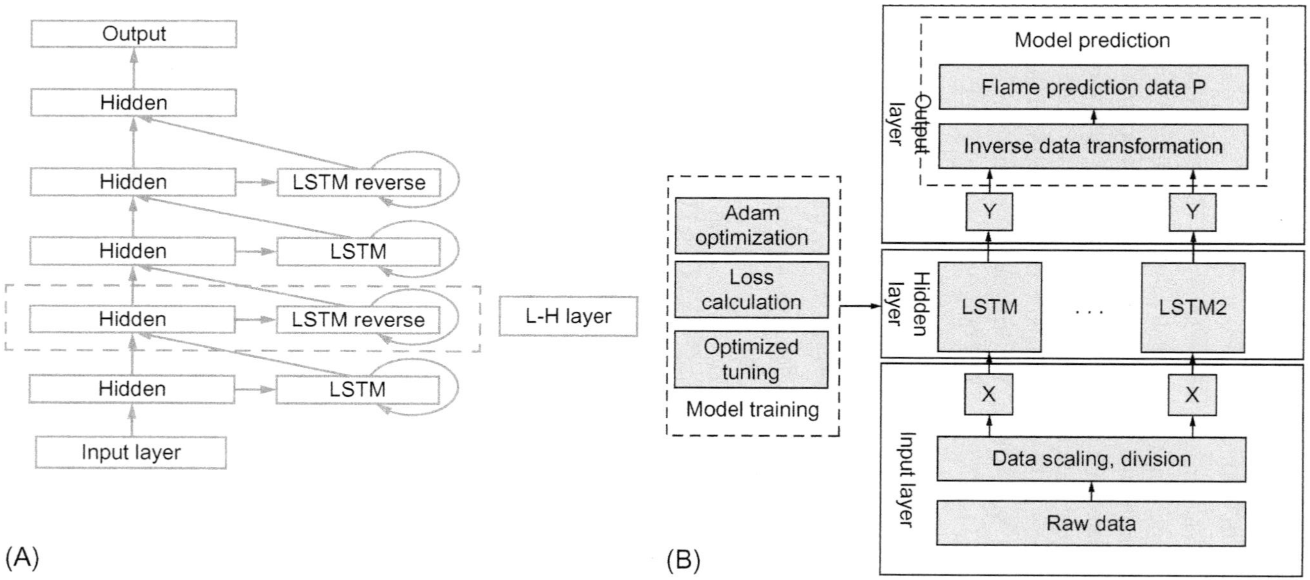

(A) (B)

FIG. 6.4

(A) Flowchart for training the LSTM. (B) Framework for LSTM combustion quality prediction model based on flame shape.

a) Adapted from http://fastml.com/deep-learning-architecture-diagrams/ b) Redrawn from V. John, S. Mita, H. Tehrani, K. Ishimaru, Automated driving by monocular camera using deep mixture of experts, in: 2017 IEEE Intelligent Vehicles Symposium (IV), IEEE, 2017.

3. Information update:

$$C_t = f_t{}^*C_{t-1} + i_t{}^*C'_t \tag{6.4}$$

4. Information output:

$$O_t = \delta(w_o \bullet [h_{t-1}, x_t] + b_o) \tag{6.5}$$

And

$$h_t = O_t{}^* \tanh(C_t) \tag{6.6}$$

It is based on the training data to iteratively update the weight of the neural network. It combines the superior performance of the two algorithms to solve sparse gradients and noise problems. Adam's tuning is relatively simple, and setting parameters can solve most problems. The Adam algorithms dynamically adjust the learning rate for each parameter based on the first-order moment and the second-order moment estimation of the gradient of each parameter according to the loss function.

6.4.3 Model optimization metrics

This chapter evaluates the performance of the model from three aspects: (1) accuracy, (2) fitting effect, and (3) running time. The mean square error (MSE) is used as the evaluation index of the model accuracy. The formula is as follows:

$$\mathrm{MSE} = \frac{1}{n}\sum_{i=1}^{n}(f_i - y_i)^2 \tag{6.7}$$

where f_i and y_i are the predicted value and the true value, respectively. For the fitting effect, we compare test results and real results of each optimization model; the fitness is the evaluation index of the model. For running time, we count the running time of each optimization model during the training test to evaluate the model's computational efficiency.

6.5 Objective of the work

The major objective of this work is to infer flue gas emissions (CO, CO_2, NO_x, and SO_x) and combustion quality from the shape and color of the furnace flame using LSTM, so that the entire firing process can be automated using IoT. This scheme will enable the control engineer to monitor and control the process parameters from any place at any point of time.

6.6 Hypothesis of this work

In our research, we selected the image-processing technique to monitor CO, CO_2, NO_x, and SO_x emissions and to estimate the combustion quality in power plants. There is only limited research to refer to, however, so we have undertaken a study in order to predict and monitor combustion quality in the gas turbine plant system.

The major hypotheses of this investigation are as follows:

- The combustion quality can be estimated from the intensity of the flame images.
- The color of the furnace flame categorizes the combustion process as complete or incomplete.
- The color of the furnace flame estimates the flue gases, which are correlated with the combustion process. The core of the fire ball is yellowish-white during complete combustion.
- Under complete combustion conditions, NO_x, SO_x, CO, and CO_2 emissions are within safety limits using LSTM.
- The gas analyzer equipment can be replaced by the proposed online, IoT-based intelligent algorithms.
- The system ensures minimal flue gas emissions at the furnace level thereby ensuring complete combustion.

6.7 Experimental environment and data preparation

The experimental data comes from the flame monitoring system of a certain working face of a thermal power plant. Morphological operations like erosion and dilation are done after edge detection using the Canny operator (results are shown in Figs. 6.5–6.8A–E) for various flame conditions like too long, too long and peripherally irregular and too short and peripherally irregular, respectively. The features (Table 6.1 obtained for one flame image under each category) are extracted from a total of 204 flame images segregated into four categories. The command STAT PROPS in MATLAB is used for this purpose. In total, we collected 51 samples corresponding to each type of flame for four types of flames. The input variables include area, centroid, bounding box, major axis length, minor axis length, eccentricity, orientation, convex area, Euler number, filled area, equivalent diameter, solidity, and perimeter. The plot of the experimental data is shown as a curve in Fig. 6.9.

6.8 Experimental results and discussion

The feature set represents the length of the flame sequence that the LSTM can utilize. It is a reaction of the length of the data association. To study the effect of the time batch size on the model, we set the number of LSTM network layers to two and the number of neurons to 34. The performance of the time batch size is 10, 20, 50, and 100, respectively. By training the flame data using LSTM, we can significantly improve the accuracy of combustion quality prediction through information fusion, and thus the model converges better. The convergence is inferred based on the Mean Squared Error (MSE) value. The combustion quality prediction model with two features has poor performance at times as shown in Table 6.2. The model with multiple input variables can better fit the trend of change for LSTM as depicted in Fig. 6.10.

The model uses multiple features as a set of input variables for information fusion to improve the performance of the LSTM and RNN. To prove the efficiency of the multiple-feature set, we compare the results of the two-feature set in Fig. 6.10A and

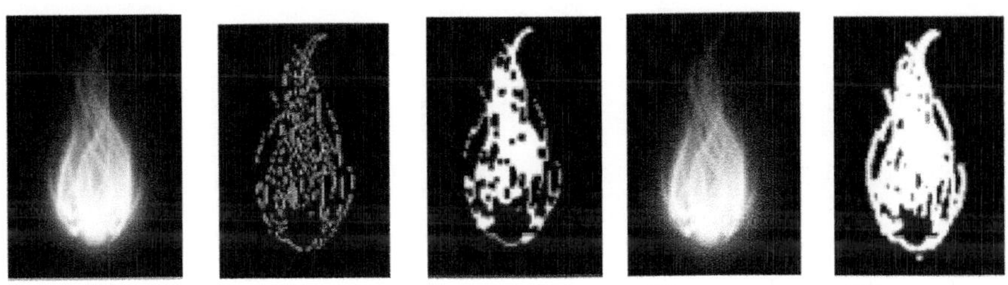

(A) (B) (C) (D) (E)

FIG. 6.5

Results for edge detection and morphological operations—appropriate flame shape. (A) Original flame image, (B) canny edge image, (C) eroded image, (D) gray scale image, and (E) Dilated image.

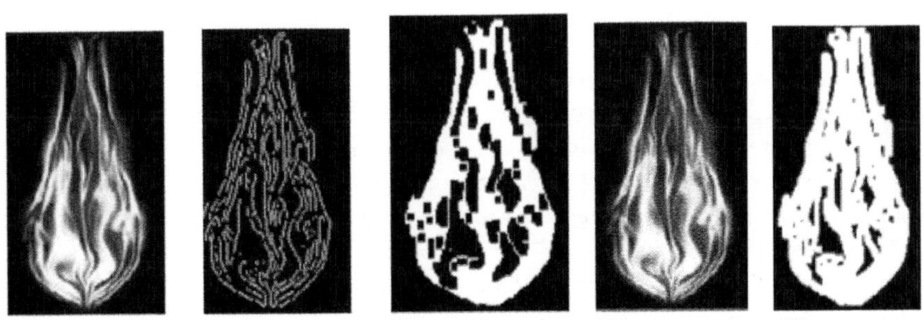

(A) (B) (C) (D) (E)

FIG. 6.6

Results for edge detection and morphological operations—flame too long. (A) Original flame image, (B) canny edge image, (C) eroded image, (D) gray scale image, and (E) Dilated image.

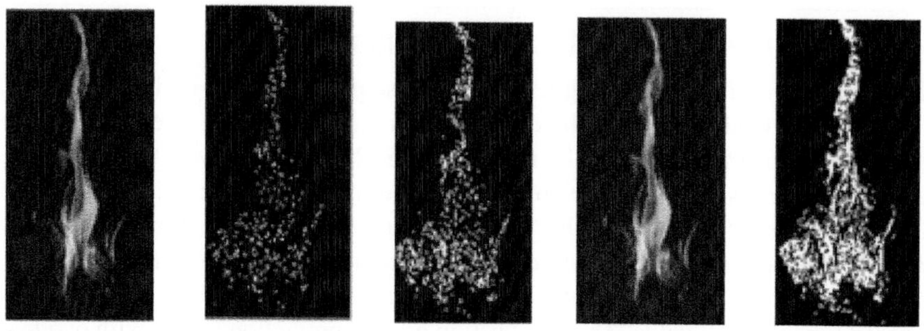

FIG. 6.7

Results for edge detection and morphological operations—flame too long and peripherally irregular. (A) Original flame image, (B) canny edge image, (C) eroded image, (D) gray scale image, and (E) Dilated image.

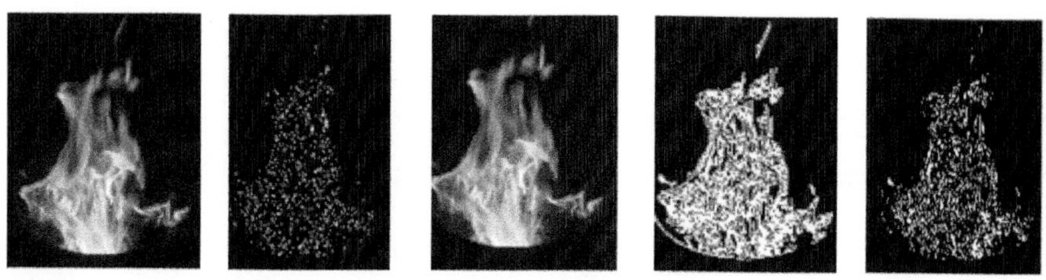

FIG. 6.8

Results for edge detection and morphological operations—flame too short and peripherally irregular. (A) Original flame image, (B) canny edge image, (C) eroded image, (D) gray scale image, and (E) Dilated image.

Table 6.1 Feature extraction

Properties	Appropriate flame shape	Flame too long	Flame too long and peripherally irregular	Flame too short and peripherally irregular
Area	289,811	579,623	1,159,245	5,629,801
Centroid	[155.977 117.083]	[311.955 234.166]	[623.9115 468.3331]	[972.5008 1.4485e+003]
Bounding Box	[0.5000 0.5000 312,467]	[0.5000 0.5000 623,934]	[0.5000 0.5000 1246 934]	[0.5000 0.5000 1944 2896]
Sub array	{[1 × 312 double] [1 × 467 double]}	{[1 × 623 double] [1 × 934 double]}	{[1 × 934 double] [1 × 1246 double]}	{[1 × 2896 double] [1 × 1944 double]}

Table 6.1 Feature extraction *Continued*

Properties	Appropriate flame shape	Flame too long	Flame too long and peripherally irregular	Flame too short and peripherally irregular
Major Axis Length	0.359425 e^{+003}	0.71885 e^{+003}	1.4377e^{+003}	3.3440e^{+003}
Minor Axis Length	0.269475 e^{+003}	0.53895 e^{+003}	1.0779e^{+003}	2.2447e^{+003}
Eccentricity	0.2439	0.5822	0.6617	0.7412
Orientation	54.3286	7.3289	0.0687	89.9999
Convex Hull	[9 × 2 double]	[9 × 2 double]	[9 × 2 double]	[9 × 2 double]
Convex Image	[312 × 467 logical double]	[623 × 934 logical double]	[934 × 1246 logical]	[2896 × 1944 logical]
Convex Area	290,941	581,882	1,163,764	5,629,824
Image	[312 × 467 logical]	[623 × 934 logical]	[934 × 1246 logical]	[2896 × 1944 logical]
Filled Image	[312 × 467 logical]	[623 × 934 logical]	[934 × 1246 logical]	[2896 × 1944 logical]
Filled Area	290,928	581,855	1,163,710	5,629,824
Euler Number	11	1475	2950	22
Extrema	[8 × 2 double]	[8 × 2 double]	[8 × 2 double]	[8 × 2 double]
Equivalent Diameter	0.303725 e^{+003}	0.60745 e^{+003}	1.2149e^{+003}	2.6773e^{+003}
Solidity	0.9883	0.9922	0.9961	1.0000
Extent	0.9883	0.9922	0.9961	1.0000
Pixel Idx List	[289,811 × 1 double]	[579,623 × 1 double]	[1159245 × 1 double]	[5629801 × 1 double]
Pixel List	[289,811 × 2 double]	[579,623 × 2 double]	[1159245 × 2 double]	[5629801 × 2 double]
Perimeter	4838	2.2026 e^{+003}	4.4052e^{+003}	9676

B using LSTM and RNN. The lowest values of the MSE are achieved while training the LSTM as compared with RNN. Table 6.2 denotes that though the time taken for training the LSTM is slightly high, it obtains the lowest values of the MSE as compared to RNN. The key to the combustion prediction model is the identification of flame shape. Therefore we study the prediction of flame shape for the next time period. The predicted length of the model is closely related to the selected feature set. The predicted length of the time step is equivalent to the length of the feature set. According to the previous experimental data, we choose the optimal feature set of

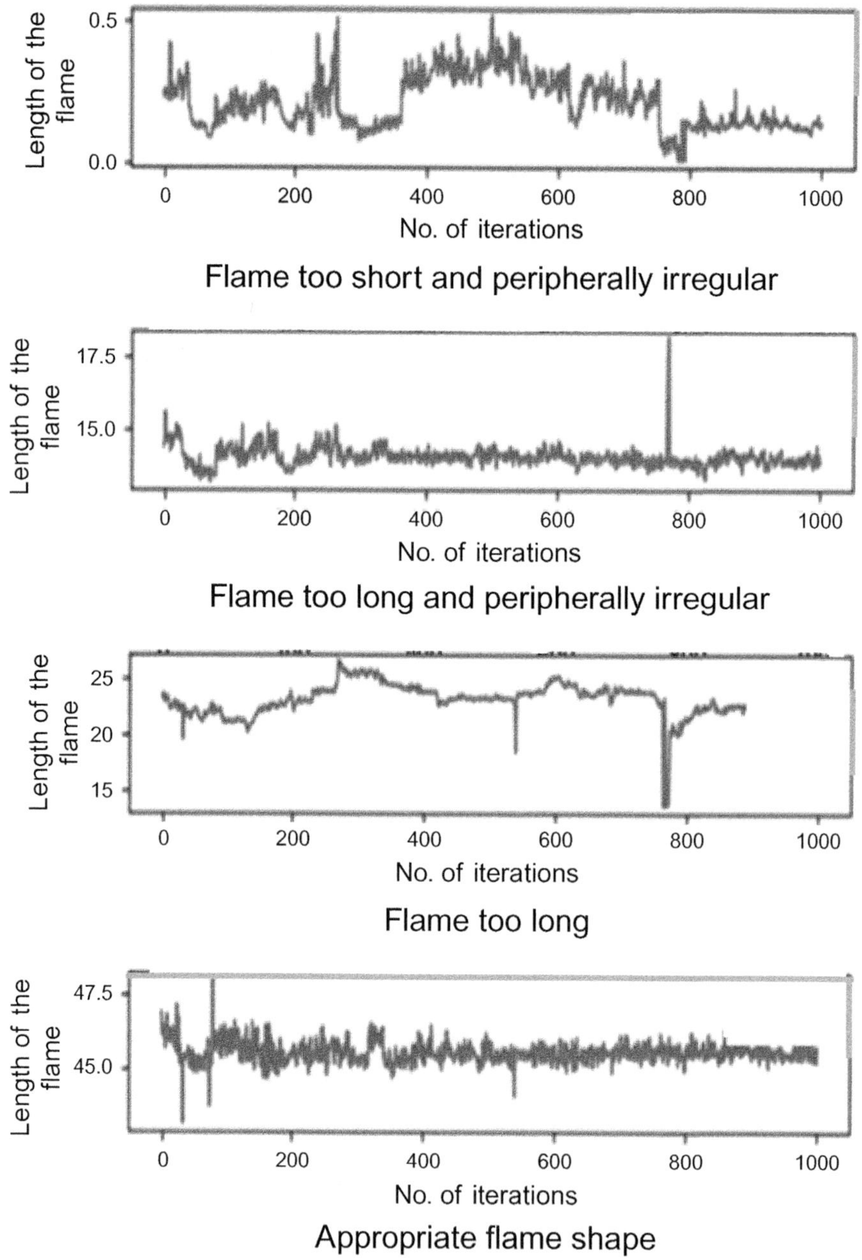

FIG. 6.9

Experimental data based on flame features.

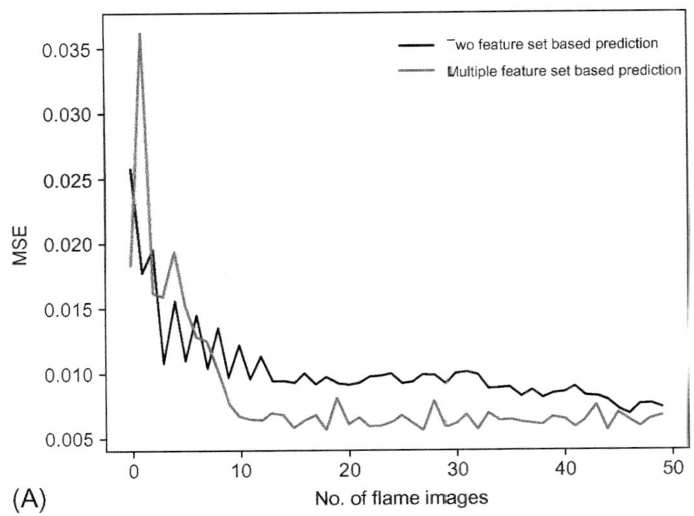

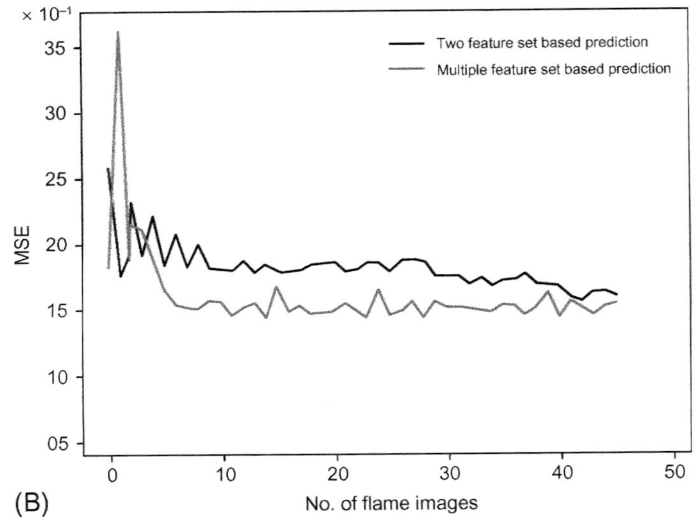

FIG. 6.10

(A) Flame feature set prediction for training the LSTM and (B) Flame feature set prediction for training the RNN.

13 from 51 flame images under four categories. Fig. 6.11 shows the graph for the error values during the LSTM training and testing process. The testing graph matches the training graph with small variations. Comparison of the prediction results for flame shape is shown in Fig. 6.12, the inference for which is recorded in Table 6.3. The training and testing parameters are recorded in Table 6.4.

According to the experimental conclusion, flame shape determines combustion quality, which is predicted by the amount of related information between the time-series data that the LSTM can utilize. The number of network layers affects the learning ability, training time, and test time of the model. A set of multiple features can improve the performance of the model. The feature set contains 13 features for 51 flame images under each category (four categories), a two-layer LSTM network results, using multiple features as the parameters of the prediction model, the prediction length is 51 unit length. Fig. 6.13 shows that within a

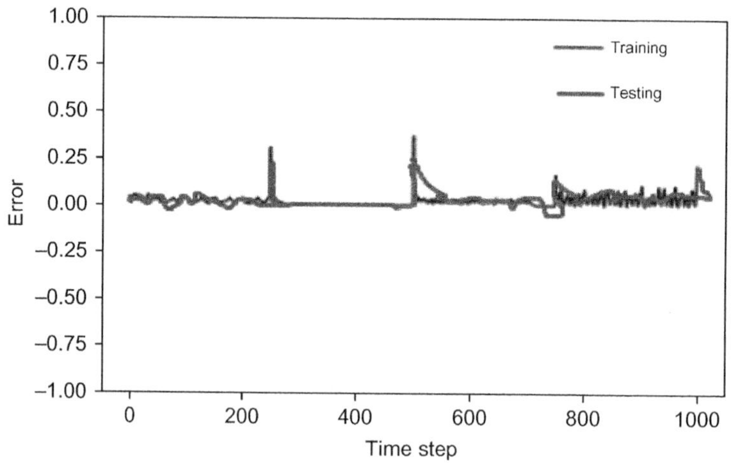

FIG. 6.11

Error graph for training and testing of LSTM.

Table 6.2 Optimizing the feature set

LSTM/RNN is two layers with 51 flame images and 34 neurons						
	Time (sec)		Difference in duration (sec)	MSE (no units)		Difference in MSE
Feature	LSTM	RNN		LSTM	RNN	
Two-feature set	2.7	2.3	0.4	0.010135	0.346	0.335865
Multiple-feature set	3	2.8	0.2	0.008331	0.125	0.116669

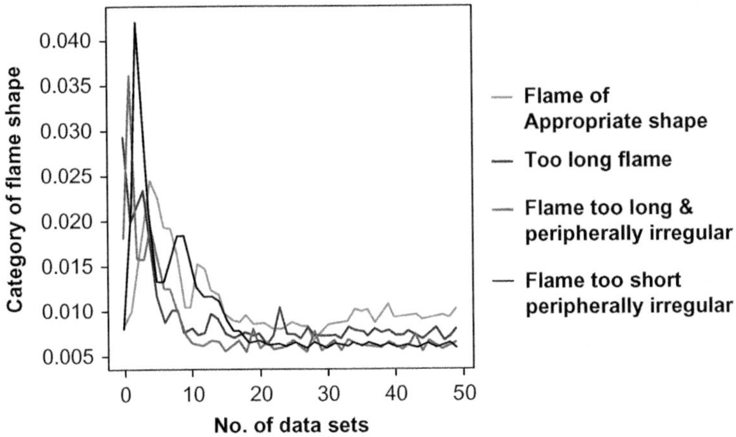

FIG. 6.12

Flame shape inference for combustion quality estimation using LSTM.

Table 6.3 Performance measure for flame shape-based combustion quality estimation

LSTM/RNN is two layers with 34 neurons				
	Time (sec)		MSE (no units)	
Flame images	LSTM	RNN	LSTM	RNN
10	9	8	0.010897	0.20972
20	7	10	0.009293	0.08972
50	4	5	0.008331	0.07322
100	2	4	0.009689	0.06827

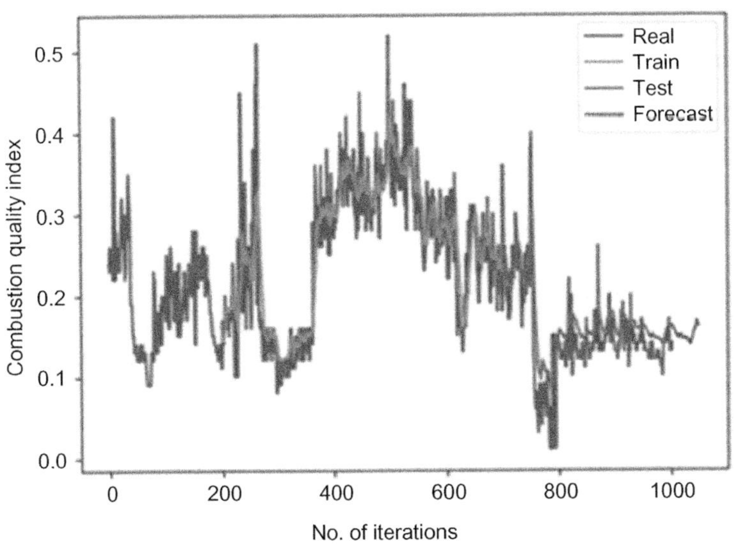

FIG. 6.13

Prediction of combustion quality index based on flame shape by LSTM during testing, training, and validation.

Table 6.4 Parameters for LSTM/RNN during training/testing

S. No	Training parameters	Values for LSTM	Values for RNN
1.	Train-time-step	3928	4300
2.	Learning rate	0.08	0.5
3.	Momentum	0.85	0.6
4.	Training RMSE	0.0918	0.3535
5.	Test RMSE	0.0823	0.2348
6.	Activation function	Gaussian	Gaussian
7.	No. of nodes in Input layer	22	22
8.	No. of hidden layers and nodes	12	15
9.	No. of nodes in output layer	04	04

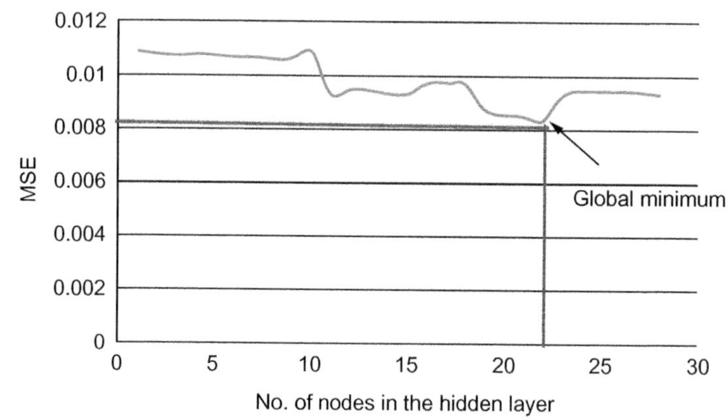

FIG. 6.14

Determination of number of nodes in the hidden layer.

certain prediction range, the model can predict the concentration of gas with high accuracy standards. The number of nodes in the hidden layer is fixed based on the lowest value of the MSE achieved as in Fig. 6.14. Thus the model can be well learned to change the combustion quality after training, especially at the time inflection point of flame shape change, which can show the superiority of the model.

As shown in Fig. 6.15, the LSTM model, after training, can diagnose combustion quality based on flame shape. It works particularly well during incomplete combustion.

The power plant turbine is the core element in power generation. Malfunctioning in the turbine system might decrease the efficiency of the plant, which leads to damage and failure of the system. This increases costs and causes major issues of combustion quality, which ultimately leads to pollution in the environment. Therefore a turbine monitoring system is needed to avoid damage caused by indecorous combustion in the plant system. As such, our focus is on monitoring and control in the combustion unit of the turbine using image processing with associated AI techniques.

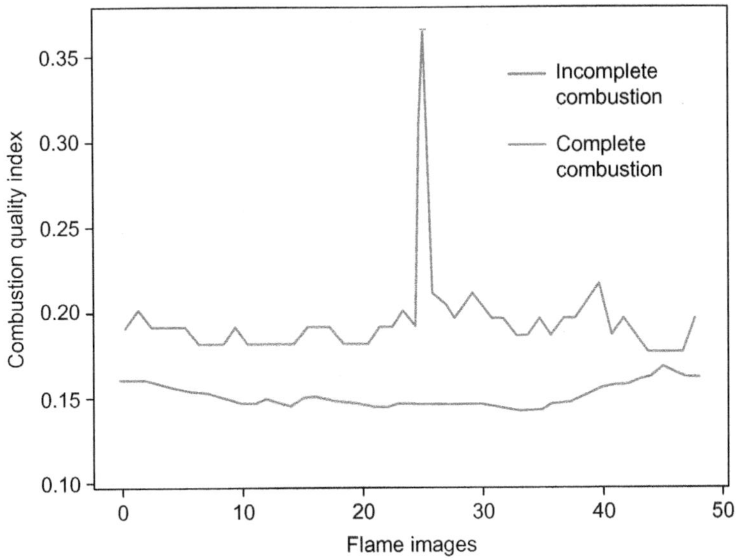

FIG. 6.15

Combustion quality prediction using LSTM for complete and incomplete combustion.

At the initial stage, we collected flame images pertaining to three different combustion conditions: (1) complete, (2) partial, and (3) incomplete combustion. However, the intermediate combustion conditions of less partial, highly partial, less incomplete, and highly incomplete were added to the data in order understand the combustion conditions and fuel emission analysis in a more efficient manner. We gathered 102 flame images from the control room for a boiler in the power station; 49 images were identified to be correct.

The previous results support that the parallel architecture of the intelligent classifiers are beneficial for combustion quality monitoring in power stations; the training and testing results are very close to the validation results. It was found that PRBFBPA gives maximum classification performance when compared to FLD and RBF. However, the classification performance could be improved by preprocessing the acquired images. The necessary action to be taken to increase or decrease the air supply which subsequently ensures the complete combustion. Our work can be further extended by considering the spectrum of the flame images. In order to avoid error and allow complete combustion by monitoring the process, another defined classifier is needed, thus we included the feature extraction process for each specific image with selected parameter.

Continuing with feature extraction, we further classified the output of edge detection using the FNN method in order to estimate NO_x and SO_x emissions and combustion quality. We used best predictive analytics software (7 features and 10 features) and implemented ant colony optimization (ACO) to improve the detection protocol. We used the same sample of flame images in this classification. ACO performances showed slight variation in precision and recall, thus proving that it is better for monitoring and controlling the combustion process. Also, it may subsequently reduce the emission of NO_x and SO_x gases to the environment. As a part of our work plan, we installed an intelligent system to monitor and control the air-fuel ratio. To control air pollution, flue gas emissions need to be reduced. The major idea behind this research is to identify adverse combustion conditions, to provide a number of quantifiable parameters to evaluate flame stability and combustion, and to develop a PC-based system for flame monitoring in thermal power plants.

6.9 Conclusion

In the investigation of combustion, quality of flame images is of great importance for detecting, recognizing, and understanding combustion conditions. Our proposed system uses soft sensors integrated with the IoT to monitor and control combustion quality, flame temperature, and CO, CO_2, NO_x, and SO_x emissions in thermal and gas power plants.

In this chapter, we proposed a prediction model for combustion quality-based flame shape using LSTM. The optimal model is obtained by training, predicting, and tuning the LSTM model. Compared with traditional methods of predicting combustion quality, LSTM technology proved that it can predict combustion quality at the next moment based on flame shape.

During the training process, selecting feature size as an input to the LSTM layer has a great influence on the objective function value, fitting effect, and running time. The appropriate feature size and number of layers can effectively improve the accuracy of the model and fit the effect and reduce the training runtime.

The LSTM model can predict the combustion quality index in the next time period in a short time range, especially during complete and incomplete combustion conditions, which can better reflect the superiority of LSTM prediction time-series data.

In general, the model effectively predicts the combustion quality index in the next time period as compared with other methods. However, the technology needs to be further improved, especially in the case of low accuracy due to integration with real-time distributed control systems (DCS).

References

[1] H. Das, R.K. Barik, H. Dubey, D.S. Roy, Cloud Computing for Geospatial Big Data Analytics: Intelligent Edge, Fog and Mist Computing, Springer, 2019.

[2] K.H.K. Reddy, H. Das, D.S. Roy, A data aware scheme for scheduling big-data applications with SAVANNA Hadoop, in: Futures of Network, CRC Press, 2017.
[3] B.S.P. Mishra, H. Das, S. Dehuri, A.K. Jagadev, Cloud Computing for Optimization: Foundations, Applications, and Challenges, vol. 39, Springer, 2018.
[4] C. Pradhan, H. Das, B. Naik, N. Dey, Handbook of Research on Information Security in Biomedical Signal Processing, IGI Global, Hershey, PA, 2018, pp. 1–414.
[5] P.K. Pattnaik, S.S. Rautaray, H. Das, J. Nayak (Eds.), Progress in computing, analytics and networking, Proceedings of ICCAN 2017, In: vol. 710, Springer, 2018.
[6] C.R. Panigrahi, M. Tiwary, B. Pati, H. Das, Big data and cyber foraging: future scope and challenges, in: Techniques and Environments for Big Data Analysis, Springer, Cham, 2016, pp. 75–100.
[7] N.P.G. Bhavani, K. Sujatha, Soft sensor for temperature measurement in gas turbine power plant, Int. J. Appl. Eng. Res. (2014) 21305–21316.
[8] G. Lu, Y. Yan, S. Cornwell, M. Whitehouse, G. Riley, Impact of co-firing coal and biomass on flame characteristics and stability, Fuel 87 (2007) 1133–1140.
[9] G.L. G Gilabert, Y. Yan, Three dimensional visualisation and reconstruction of the luminosity distribution of a flame using digital imaging techniques, in: Journal of Physics: Conference Series – Sensors and their Applications XIII. Journal of Physics: Conference Series, 15 2005, pp. 167–171.
[10] G. Lu, Y. Yan, Advanced monitoring and characterization of combustion flames, in: 20th Annual Meeting and Meeting of the Advanced Power Generation Division, University of Leeds, 2009.
[11] K. Sujatha, N. Pappa, Combustion monitoring of a water tube boiler using a discriminant radial basis network, ISA Trans. 50 (2011) 101–110.
[12] K. Sujatha, N. Pappa, Combustion quality estimation in Power Station boilers using median threshold clustering algorithms, Int. J. Eng. Sci. Technol. 2 (7) (2010) 2623–2631.
[13] K. Sujatha, N. Pappa, K. Senthil Kumar, U. Siddharth Nambi, Monitoring power station boilers using ANN and image processing, Adv. Mater. Res. 631-632 (2013) 1154–1159.
[14] K. Sujatha, N. Pappa, K. Senthil Kumar, U. SiddharthNambi, C.R. Raja Dinakaran, Intelligent parallel networks for combustion quality monitoring in Power Station boilers, Adv. Mater. Res. 699 (2013) 893–899.
[15] K. Sujatha, N. Pappa, K. SenthilKumar, U. SiddharthNambi, C.R. Raja Dinakaran, Automation of combustion monitoring in boilers using discriminant radial basis network, Int. J. Artif. Intell. Soft Comput. 3 (2013) 257–275.
[16] K. Sujatha, N. Pappa, Flame monitoring in power station boilers using image processing, ICTACT J. Image Video Process. 2 (4) (2012) 427–434.
[17] K. Sujatha, U. SiddharthNambi, Combustion quality estimation in power station boilers using SVM based feature reduction with Bayesian, Eur. J. Sci. Res. 120 (2014) 189–198.
[18] K. Sujatha, T. Senthil Kumar, R. Godhavari, S. Ponmagal, N.P.G. Bhavani, An effective distributed service model for image based combustion quality monitoring and estimation in power station boilers, in: H. Sulaiman, M. Othman, M. Othman, Y. Rahim, N. Pee (Eds.), Advanced Computer and Communication Engineering Technology, Lecture Notes in Electrical Engineering (LNEE), vol. 362, Springer, Cham, 2016, pp. 33–49.
[19] K. Sujatha, N. Pappa, Combustion quality monitoring in power station boilers using SVM based feature reduction and RBF, in: Trends in Industrial Measurements and Automation, 2011.

[20] K. Sujatha, Flame image analysis for combustion quality estimation and power station boilers using classification algorithms, in: International Conference on Sustainable Energy and Intelligent System, 2011.
[21] C.K. Tan, S.J. Wilcox, J. Ward, M. Lewitt, Monitoring near burner slag deposition with a hybrid neural network system, Meas. Sci. Technol. 14 (2003) 1137.
[22] G. Sabetian, F. Jahromy, H.M. Kamrava, Flame detectors in power plants, Int. Sci. Invest. J. 3 (1) (2014) 99–107.
[23] M. Shakil, M. Elshafei, M.A. Habib, F.A. Maleki, Soft sensor ForNOx and O_2 using dynamic neural networks, Comput. Electr. Eng. 35 (4) (2009) 578–586.
[24] K. Sujatha, N. Pappa, Combustion quality monitoring in PS boilers using discriminant RBF, ISA Trans. 2 (7) (2011) 2623–2631.

CHAPTER

Communication-aware edge-centric knowledge dissemination in edge computing environments

7

Stefanos Nikolaou, Christos Anagnostopoulos, Dimitrios Pezaros
School of Computing Science, University of Glasgow, Glasgow, United Kingdom

7.1 Introduction

7.1.1 Problem description

With the introduction of Internet Protocol version 6 (IPv6) more and more devices can be connected to the Internet, bringing the Internet of things (IoT) to people's day-to-day life [1]. IoT is enabling the use of large sensor networks [2], which can he IPv6 uses a 128-bit address space, which can theoretically provide 2^{128} unique IP addresses. IPv4 uses a 32-bit address space, which can provide 2^{32} unique IP addresses. The new IP protocol can provide more than 340 undecillion new IP addresses for potentially all the electronic devices in the world, thus creating a need for time-consuming and highly expensive investment in the communications infrastructure. Even if new communication infrastructure becomes immediately available, this exponential increase in its use will create a huge power consumption, which will burden our energy footprint.

7.1.2 Contributions and assumptions

This literature proposes an edge-centric methodology based on cached models, where communication overhead is considerably reduced. This methodology contains a mechanism that can automatically decide when to update and transmit the cached models. This chapter provides the experimental design used in this research. The experiment is designed and implemented in a way to provide useful information in order to be able to compare the cached models. Moreover, we present the comparative results of the experiment's outcome and perform further analysis to rank the models based on their performance in terms of accuracy and network efficiency.

It is assumed that the data used with the proposed methodology are not chaotic, thus it is possible for these values to be modeled and accurately predicted, for example, natural values like temperature or humidity.

Real-Time Data Analytics for Large Scale Sensor Data. https://doi.org/10.1016/B978-0-12-818014-3.00007-3

7.1.3 Report structure

This chapter has five sections. Section 7.1 is the introduction where we present the problem driving this research as well as a brief description of the methodology used and the assumptions made concerning the data. Section 7.2 provides the literature review. Section 7.3 presents our rationale and provides fundamental definitions. Section 7.4 describes the proposed methodology and Section 7.5 describes the experimental design. The experimental design contains information including system elements like end nodes and base nodes, the network topology and network parameters, the dataset, the requirements, the methodology and the proposed experiment. Section 7.6 provides results and presents them in the form of graphs, text, and tables. Finally, Section 7.7 concludes our research and propose future work to implement.

7.2 Literature review

In the current literature the benefits of using edge-computing to lower computational overhead has been demonstrated [3]. Harth and Anagnostopoulos [3] introduced a communication-efficient method that transmits only regression model parameters and sufficient statistics for cached models when certain conditions occur. These conditions are referred to in the literature as optimal stopping rules and are part of optimal stopping theory. Panagidi et al. [4] studied the use of optimal stopping theory to find an optimal stopping rule in order to ensure the optimal delivery of important information, such as control messages and telemetry. Pasteris et al. [5] proposed a methodology for data distribution over computing devices. This methodology downgrades the bandwidth used and eliminates the storage constraints. Kolomvatsos and Anagnostopoulos [6] studied a methodology that pushes the analytical tasks to the network edge. In their proposed methodology, the nodes use their resources to execute these tasks, but due to nodes' resource constraints they are able only to execute a limited number of tasks. To overcome this the nodes select only the tasks that will maximize performance. The rest are processed either on neighboring networking peer nodes or at the cloud. The proposed method efficiently decides where each task is going to be allocated for execution. This methodology significantly saves communication overhead because in the vast majority of cases, the tasks are processed locally on neighboring nodes. Kontos et al. [7] discussed the input of the optimal stopping problem in optimizing the scheduling of transmission in noisy epidemic dissemination environments. The main purpose of this research is to achieve significant energy cost reduction by scheduling the transmission times. Many papers discuss the use of optimal stopping theory to discover an optimal stopping point or a set of rules (as described in Ref. [4,7]). However, in our research we use a slightly different approach in order for the models to have fair comparison. Furthermore in this work we build upon the mechanism described by Harth and Anagnostopoulos [3], with the following key innovations:

We provide an analysis to determine which regression models combined with a similar methodology as the one proposed by Mishra et al. [2] can provide the best trade-off between downgrading the computational overhead while having a small toll in accuracy.

We describe what happens in terms of accuracy and communication overhead to this methodology with different regression models, such as cached models. Evidently, there will always be a trade-off between accuracy and communicational overhead. However, by selectively transmitting we can minimize communication overhead such that the energy consumption in edge nodes, network elements, and base nodes will be decreased. Additionally, more and more bandwidth will become available, which can result in better communications [8].

7.3 Rationale and fundamentals

7.3.1 Rationale

The rationale behind the proposed methodology is to decrease the communication overhead between an end node and a base node in large networks (networks with many entities), such as sensor networks [2]. Such downgrade in the communication overhead can be achieved by using model caching methodologies. For instance, if an end node transmits a sensing value without significant change over time, the quality of the analytics in the base node will not change, even if the new values are missing. Traditional communications will continuously transmit the data compared to the proposed methodology, which will not send the new parameters of the model. This is due to the fact that the cached model has not significantly changed over time. Furthermore continuous transition of data could result in a bottleneck, where a big delay could result in low analytics quality in the base node. Moreover, there are applications where the end nodes need to minimize their communication transactions. One of these applications is weather monitoring in remote areas. In situations like these, the sensors need to minimize their communication transactions with their base node in order to save energy. By saving energy they allow the battery of this sensing device to last longer, so the research can be conducted uninterrupted for longer periods of time.

7.3.2 Definitions

Median absolute error (MedAE): The MedAE is the median of all the absolute differences between the predicted values and the real values. This metric is particularly robust in outliers, because extreme values does not affect the outcome of this metric.

$$\mathbf{MedAE}(y,\hat{y}) = \mathbf{median}(|y_1 - \hat{y}_1|, \ldots, |y_n - \hat{y}_n|) \tag{7.1}$$

In this research we used the MedAE (Eq. 7.1) because it is not easily altered by outliers. With the use of this metric, the proposed methodology can provide a robust solution against incoming outliers on how to find significant changes between an updated model and the previous cached model in the memory.

Mean square error (MSE): The MSE is the mean squared difference of the predicted values and the real values. This metric penalizes bigger differences, which is useful when comparing prediction performance of various models.

$$\mathbf{MSE}(y,\hat{y}) = \frac{1}{n}\sum_{t=1}^{n}(y_t - \hat{y}_t)^2 \tag{7.2}$$

In Section 7.5, the MSE is used to compare the accuracy of the models. Because the errors are squared before they are averaged, this metric gives an exponential higher weight in bigger errors. This makes it possible to more easily distinguish which is the best performing model when outliers do not exist. When the models are tested for their accuracy in predictions, outliers are not present.

Linear regression (LR): Eq. (7.3) is the equation of the linear regression (LR) model in which $x_n$is the input vector and w is the vector of weights.

$$y_n = w^T x_n \tag{7.3}$$

The loss function of LR models used in this literature is the mean square deviation (Eq. 7.4):

$$L = \frac{1}{N}(\mathbf{y} - \mathbf{Xw})^T(\mathbf{y} - \mathbf{Xw}) \tag{7.4}$$

In order to predict y_n with high accuracy we need to calculate the vector w. This vector is calculated by equating the first partial derivative of the loss function of y_n (Eq. 7.5) with zero and solve it for w.

$$\frac{\partial L}{\partial w} = 0 \tag{7.5}$$

After solving the Eq. (7.5) the vector w is:

$$w = \left(X^T X\right)^{-1} X^T y \tag{7.6}$$

Radial-basis function (RBF) kernel: The RBF kernel (Eq. 7.7) is used as the covariance function in the Gaussian process regression (GPR). It is parameterized by the length-scale parameter l, $l \in \mathbb{R}^+$. The length-scale l indicates how smooth a function is. If $l \ll$ then the values of the function can change rapidly. If $l \gg$ the function is smooth and the values of the function cannot change significantly.

$$k\left(x^{(i)}, x^{(j)}\right) = \sigma^2 \exp\left(-\frac{1}{2}\sum_{d=1}^{n}\left(x_d^{(i)} - x_d^{(j)}\right)^2 / l_d^2\right) \tag{7.7}$$

In order to choose the hyperparameters $\theta = \{\sigma, 1\}$ of the RBF kernel, we infer θ using Bayesian methods. Bayesian methods are more costly than finding the θ by optimizing the marginal likelihood of the GPR, however, by using Bayesian methods our method is immune to overfitting.

Gaussian Process Regression (GPR): In order to be able to perform GPR we need to calculate the matrix K (Eq. 7.8). This matrix contains all the results of the covariance function applied in all the possible combinations of the records in dataset [9,10].

$$K = \begin{bmatrix} k(x_1,x_1) & k(x_1,x_2) & \cdots & k(x_1,x_n) \\ k(x_2,x_1) & k(x_2,x_2) & \ddots & k(x_2,x_n) \\ \vdots & & \ddots & \vdots \\ k(x_n,x_1) & k(x_n,x_2) & \cdots & k(x_n,x_n) \end{bmatrix} \tag{7.8}$$

In Gaussian process modeling, the data are represented as a sample from a multivariate Gaussian distribution (Eq. 7.9).

$$\begin{bmatrix} y \\ y_* \end{bmatrix} \sim N\left(0, \begin{bmatrix} K & K_*^T \\ K_* & K_{**} \end{bmatrix}\right) \tag{7.9}$$

From Eq. (7.9), we can derive the conditional probability given in the next equation:

$$p(y_*|y) \sim N\left(K_*K^{-1}y, K_{**} - K_*K^{-1}K_*^T\right) \tag{7.10}$$

The best estimate for y_* is the mean of this distribution:

$$\overline{y_*} = K_*K^{-1}y \tag{7.11}$$

The uncertainty of this estimate is captured by the variance of this distribution:

$$\mathrm{var}(y_*) = K_{**} - K_*K^{-1}K_*^T \tag{7.12}$$

7.4 Methodology

In this section the proposed methodology is described, in both physical language and pseudocode.

7.4.1 Models used with the methodology

Two regression models were used in this research. The first one is the classic LR model and the second one is the more robust GPR with RBF as kernel. Both models, as well as the kernel, are described in Section 7.3.2.

7.4.2 Algorithm description

This proposed methodology was implemented mainly for end node sensing devices. These devices have a specific period that will read/sense a value (e.g., temperature). In this research we require that these end nodes will have at least the storage and processing power of a modern micro-controller (MCU). The analysis begins when the end node collects readings and fills a window of specified size in its flash memory. When that window Y is full, our proposed methodology begins by training a model $f_0(X)$, with the window Y_0, and after that storing it in memory as $f_{\mathrm{CachedModel}}(X_0)$. Each time the end node does a new reading, the new value will be added to the window Y_i, while the oldest value will be removed from it, in order to avoid memory overflow. Furthermore when a new value is added to the window Y_i, a new model $f_i(X)$ is trained with it and after training it, both the cached model $f_{\text{CachedModel}}(X)$ as well as the $f_i(X)$ are giving the prediction $\text{Pr}_{\text{CachedModel}} = f_{\text{CachedModel}}(Y_i)$ and $\text{Pr}_i = f_i(Y_i)$. After these predictions are made, they are passed to Eq. (7.13) together with Y_i, in order for this function to calculate the model-difference between the old $f_{\text{CachedModel}}(X)$ and the newly created one $f_i(X)$. If this model-difference is above a threshold β, then the end node sends the parameters of the $f_i(X)$ to the base node and replaces the old model in cache with the newly created one.

$$\textbf{ModelDiff}(Y_i, \text{Pr}_i, \text{Pr}_{\textbf{CachedModel}}) = \frac{|\textbf{MedAE}(Y_i, \text{Pr}_i) - \textbf{MedAE}(Y_i, \text{Pr}_{\textbf{CachedModel}})|}{|\textbf{MedAE}(Y_i, \text{Pr}_i)|} \qquad Y_i \neq \text{Pr}_i \tag{7.13}$$

Algorithm 1 Algorithm for updating the model to BN

Input: sensor
Output: connect_to_BN

```
while True do
    Y.delete_last_value()
    Y.push(sensor.read_new_value())
    f_x ← train_model(Y)
    if ModelDiff(Y, f_x(Y), f_cached(Y)) > β then
        f_cached ← f_x
        connect_to_BN.send_model_params(f_x)
    end if
end while
```

7.5 Experimental design

This section presents important information about the experiments, including the network entities of the system and the data used as well as the methodology and experimental configuration.

7.5.1 Network topology

Fig. 7.1 presents the most common topology used in sensor communications. There are series of end nodes connected to a base node through a WAN or LAN and some other network components like router, firewalls, and so on.

The entities of the network are:

- *End node*: The end node can be any sensing or computing device that has at least the processing power and memory of a modern MCU, in order to be able to execute the analytical tasks needed for this research.

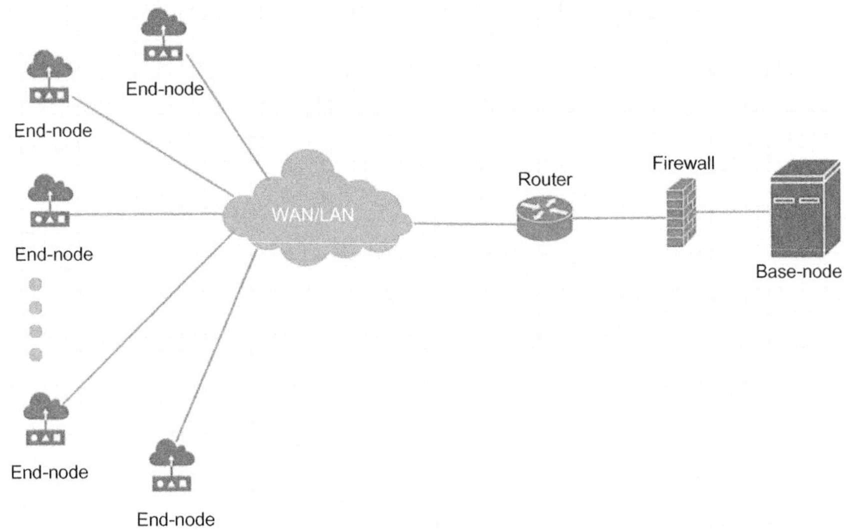

FIG. 7.1

A standard IOT network topology.

- *Base-node*: The base node is an application server, which will accept socket connections from the end nodes in order to receive the updated models.
- *LAN/WAN*: LAN can be described as a network covering a small geographical area, such as a campus, home, office, and so on. WAN can be described as a network covering large geographical areas, like region, state, nation, and so on (e.g., a country's backbone network can be described as a WAN).
- *Router*: A router is a networking device that directs the data packets traveling from one network to another and guides them to their destination based on a routing protocol following a global standard like TCP/IP.

7.5.2 Dataset

The dataset is created by the Glasgow Network Functions for Unmanned Vehicles (GNFUV) project. It comprises four csv files, each containing the data gathered from humidity and temperature sensors of each of the four unmanned surface vehicles (USVs) that took part in the GNFUV experiments. Each CSV contains 342 records from sensor readings and other vital information. After a cleaning process over these files, we created three datasets. The first dataset (Table 7.1) has only one feature, humidity, and the value to be predicted is the temperature. The second dataset's (Table 7.2) features are a window of the three previous sampled readings of temperature and the predicted value is the most recent sampled reading of the temperature. The third dataset (Table 7.3) has as features a window of the three previous samples of temperature values as well as the three previous samples of humidity values. It has the most recent sample of the temperature as the predicted value.

A small sample of each of the three datasets is shown in Tables 7.1–7.3. The *white columns* represent the index (which is the time period the data were gathered in), the *gray columns* represent the training set, and the *green column* represents the values to be predicted.

Table 7.1 This is the first dataset containing humidity as data and temperature as value to be predicted

	Humidity	**Temperature**
0	21	40
1	21	40
2	21	40
3	21	40
4	22	40
5	22	40
6	22	39
7	23	39
8	23	39
9	24	38
10	24	38
11	24	38
12	25	38
13	25	38

Table 7.2 This is the second dataset containing a window of temperature as data and temperature as values to be predicted

	Temp *t*(*t*-3)	**Temp *t*(*t*-2)**	**Temp *t*(*t*-1)**	**Temp *t*(*t*-0)**
3	40	40	40	40
4	40	40	40	40
5	40	40	40	40
6	40	40	40	39
7	40	40	39	39
8	40	39	39	39
9	39	39	39	38
10	39	39	38	38
11	39	38	38	38
12	38	38	38	38
13	38	38	38	38

Table 7.3 This is the third dataset containing a window of humidity and one of temperature as data and values of temperature as values to be predicted

	Temp (*t*-3)	Temp (*t*-2)	Temp (*t*-1)	Hum (*t*-3)	Hum (*t*-2)	Hum (*t*-1)	Temp (*t*-0)
3	40	40	40	21	21	21	40
4	40	40	40	21	21	21	40
5	40	40	40	21	21	22	40
6	40	40	40	21	22	22	39
7	40	40	39	22	22	22	39
8	40	39	39	22	22	23	39
9	39	39	39	22	23	23	38
10	39	39	38	23	23	24	38
11	39	38	38	23	24	24	38
12	38	38	38	24	24	24	38
13	38	38	38	24	24	25	38

7.5.3 Modeling techniques

In this research we use two regression models: LR and GPR with RBF kernel. Each of the aforementioned models are used with the three datasets described in the previous paragraphs thus creating a total of six modeling techniques to be tested. The first is called "linear regression (LR)" and it is LR trained on the first dataset of this research. The second is called "LR temp windowing" and it is LR trained on the second dataset of this research. The third is called "LR temp & hum windowing" and it is LR trained on the third dataset of this research. The fourth is called "Gaussian process regression (GPR)" and it is GPR with RBF kernel trained on the first dataset of this research. The fifth is called "GPR temp windowing" and it is GPR with RBF kernel trained on the second dataset of this research. The sixth is called "GPR temp & hum windowing" and it is GPR with RBF kernel trained on the third dataset of this research.

7.5.4 Experiment: Comparing the models for the proposed methodology

This experiment was designed in order to compare which modeling technique is going to be more efficient in terms of accuracy and communication overhead. The experiment was run in an ideal environment with packet-loss = 0% and latency less than 20 ms, so that the network will not be able to change the outcome of the results. All the experiments were run with the same threshold β. The value for β was selected to be 0.2, due to the fact that it creates small changes in accuracy, but big changes in communication transactions. The end nodes were using the proposed methodology to update the models to the base node. In this experiment we use network analyzers to capture the data over the media in order to collect important information about the communication overhead created by the combination of the dataset, the model, and the proposed methodology.

7.5.4.1 Experimental set-up

The network diagram in Fig. 7.2 presents the experimental set-up for gathering network information to support this literature.

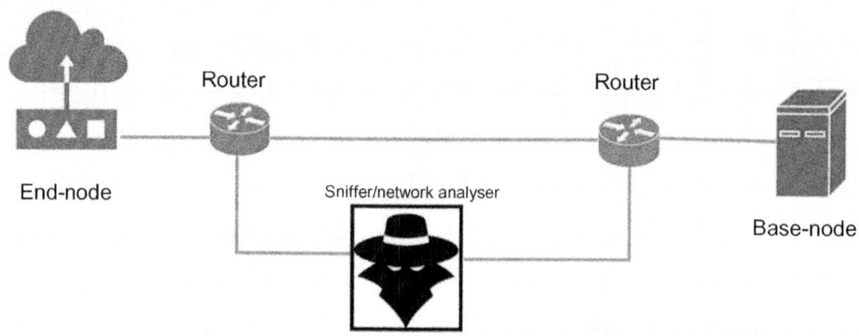

FIG. 7.2

This figure shows the network topology of the experiment. The end node is connected directly to the base node.

The entities of the network are: end node, routers, base node, and network analyzer (sniffer).

The *network analyzer* (sniffer) is a program that reads the network card of a device and logs the traffic passing through this device. In this experiment it was used in order to calculate the communication overhead created by each model.

7.5.5 Experimental process

The experiment was run in a virtual environment. The end nodes used Linux and ran on top of virtual machines (VMs) with static IPs. The static IPs were used as a way for the base node to identify each VM. Each of the end nodes ran one of the modeling techniques, so there was a total of six VMs running in this experiment. Each end node updated a model in the base node by using the methodology proposed in this research. The base node was able to match the incoming data to an end node based on the IP of the incoming packets. Then it updated the right model with the parameters contained in the received packets. Because all of the modeling techniques are based on datasets stemming from the same initial dataset, each one of their records that has the same ID number (same sensing period) represents the same set of sensed values of temperature or humidity from that original set. All the end nodes read the dataset records one by one and then ran the proposed methodology. If they decided to update their model to the base node, they sent the model parameters as well as the ID of the line that the decision of update was made. We saved all the incoming model parameters of each model to the file system of the base node, ordered by the periods in which they were sent. This allowed for a chronological order of the models, so that later we were able to access all the models the base node has received over time and thus make comparisons with each modeling technique. Finally, each end node as well as base node ran a network analyzer, so we were able to make network measurements, such as the communication overhead of an end node.

7.6 Results and evaluation

In this section we present the results from the execution of our experiment. Specifically, we present the predictions over time per model from the client side as well as from the server side. Then we present a bar chart that shows the differences between the models in terms of communication overhead. Next, we include a table of the metric of mean-absolute-error for the whole set of predictions over time of all models for the client side as well as for the server side. This table also contains the communication overhead of each model in megabytes. We normalize the three measurements in other columns of the table in order to use the ranking function to rank our models of this methodology from best (smallest value) to worst (highest value). The table also includes the ranking score given by the proposed equation.

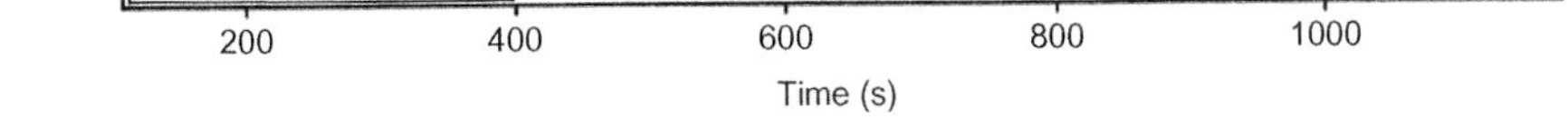

FIG. 7.3

This figure presents all the model predictions over time made in client side.

The graph in Fig. 7.3 shows the prediction of the models on the client side during the experiment. All the model predictions lie around the *blue line* (dark gray in print version), which is the real data. This happens because each time the end node reads a new value it builds a new model, so the prediction from that model will always be close to real values.

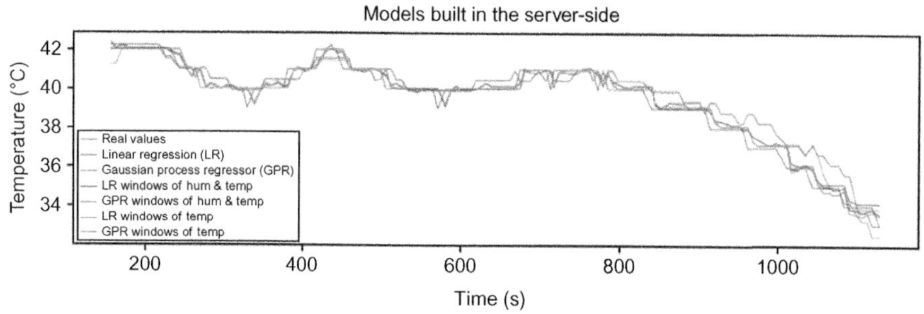

FIG. 7.4

This figure presents all the model predictions over time made in server side.

The graph in Fig. 7.4 shows the prediction of the models on the server side during the experiment. It shows that even with less communication transaction the predictions can still be very close to the real values. However, the difference between this graph (Fig. 7.4) and the graph presenting the predictions on the client side (Fig. 7.3) is considerable.

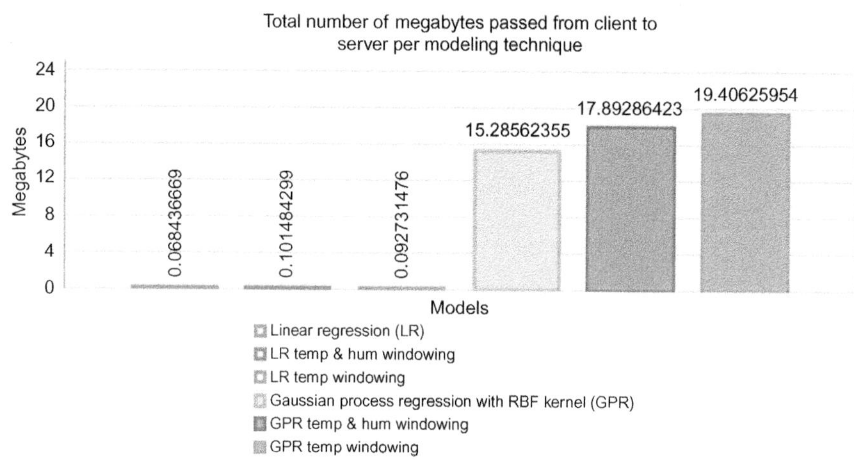

FIG. 7.5

This figure shows the communication overhead each model created to send its parameters to server.

Fig. 7.5 demonstrates the exponential difference in communication overhead between LR models and GPR models with RBF kernel.

This difference in communication overhead is due to the fact that LR transmits a dimensional array with size m containing the weights of the linear model. The GPR with RBF kernel transmits the hyperparameters of the RBF kernel as well as the covariance matrix K, which has size mxm.

Note: In this research there was not any method implemented to downsize the data sent by GPR or LR.

Table 7.4 This table contains all the metrics of each modeling technique

	MB (client- > server)	Normalized (MB (client- > server))	mean square error (client-side)	Normalized (mean square error (client-side))	mean square error (server-side)	Normalized (mean square error (server-side))	Models ranked
Linear regression (LR)	0.068	0	0.389	0.946721311	0.51	1	0.494672131
LB temp windowing	0.101	0.001706485	0.17	0.049180328	0.178	0.056818182	0.028498548
LR temp & hum windowing	0.093	0.001292791	0.158	0	0.158	0	0.000646396
Gaussian process regression (GPR)	15.285	0.786896266	0.402	1	0.453	0.838068182	0.828675406
GPR temp windowing	17.893	0.921760265	0.168	0.040983607	0.167	0.025568182	0.475205766
GPR temp & hum windowing	19.406	1	0.16	0.008196721	0.16	0.005681818	0.503092399

Table 7.4 presents all the metrics and measurements used in finding the two most efficient modeling techniques in terms of accuracy and communicational overhead. In order to be able to propose a function to rank the models we needed first to normalize all the measures/metrics. We used a classic normalization formula (Eq. 7.14).

$$\mathbf{normalize}(\mathbf{i}, \mathbf{Y_i}) = \frac{\mathbf{Y_i} - \mathbf{min}(\mathbf{Y})}{\mathbf{max}(\mathbf{Y}) - \mathbf{min}(\mathbf{Y})} \tag{7.14}$$

After the metrics/measurements where normalized we used Eq. (7.15) to rank the models. This equation uses three weights to give the appropriate importance in the normalized metrics/measurements. The first indicates that the data that traveled from client to server for each model accounts for 50% of the importance in the equation. The second states that the normalized mean-absolute-error from the client side is not very important (it is weighted only 10%), but it should be included in the formula. Finally, the third weight is considered really important because it indicates how accurate the model is on the server side.

$$\text{rank}_i(\text{table Of Metrics}) = 0.5{*}\text{table Of Metrics}(i, 1) + 0.1{*}\text{table Of Metrics} + 0.4{*}\text{tableOfMetrics}(i, 5) \tag{7.15}$$

The last column of Table 7.4 contains the ranking values of the models from best (smallest value) to worst (highest value). The smallest value of this column, and thus the best modeling technique, belongs to "LR temp & hum windowing." The second smallest belongs to "LR temp windowing."

7.6.1 Best model

Fig. 7.6 shows that the server-side model is almost the same as the client-side model. Only a few differences are visible in this graph. This means that the accuracy of the predictions at both the server and the client sides remains high even though the client sends less data to the base node. The reason behind why the features are so highly

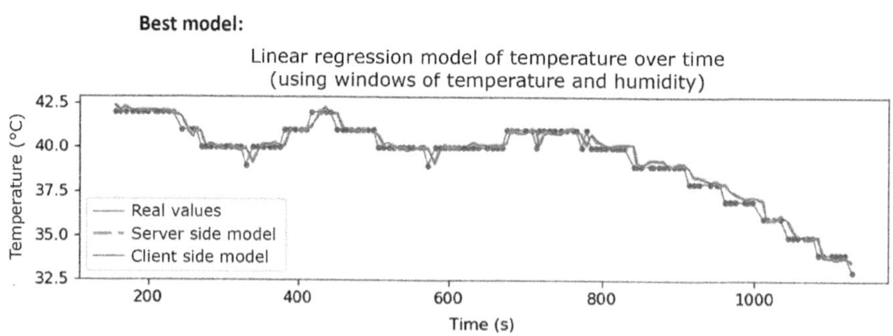

FIG. 7.6

LR model prediction in server and client sides as well as the real value (value to be predicted). In this figure the model was trained with the third dataset (Table 7.3).

correlated is based on the fact that they are the same measurements in units (e.g. the sensor read temperature) just sampled/read by the end node in previous periods.

7.6.2 Second-best model

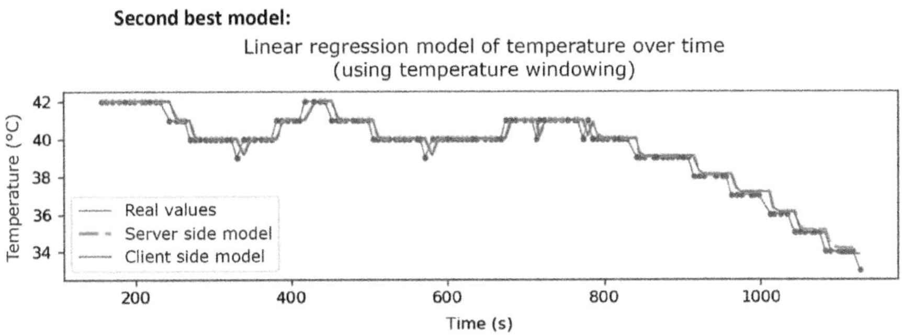

FIG. 7.7

LR model prediction in server and client sides as well as the real value (value to be predicted). In this figure the model was trained with the second dataset (Table 7.2).

Fig. 7.7 demonstrates that the server-side model is close to equal to the client-side model. As stated previously, the correlation between the features is the reason for such an accurate model.

7.6.3 The rest of the models

Similarly to the preceding, the models that follow are presented and ranked based on their performance (Figs. 7.8–7.11).

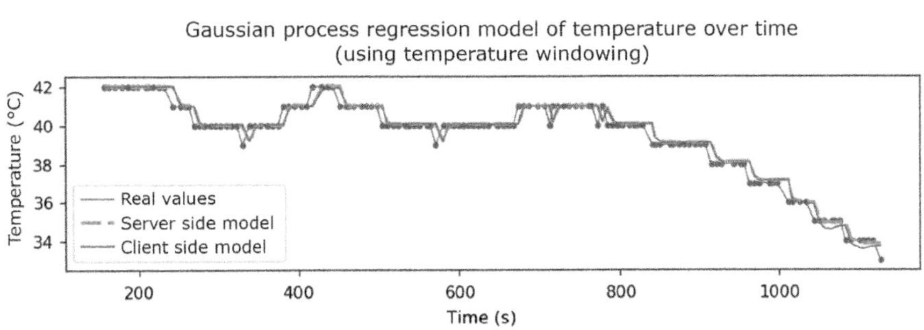

FIG. 7.8

GPR model prediction in server and client sides as well as the real value (value to be predicted). In this figure the model was trained with the second dataset (Table 7.2).

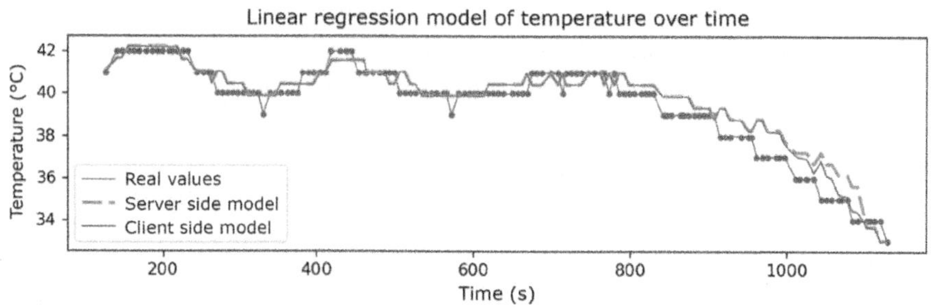

FIG. 7.9

LR model prediction in server and client sides as well as the real value (value to be predicted). In this figure the model was trained with the first dataset (Table 7.1).

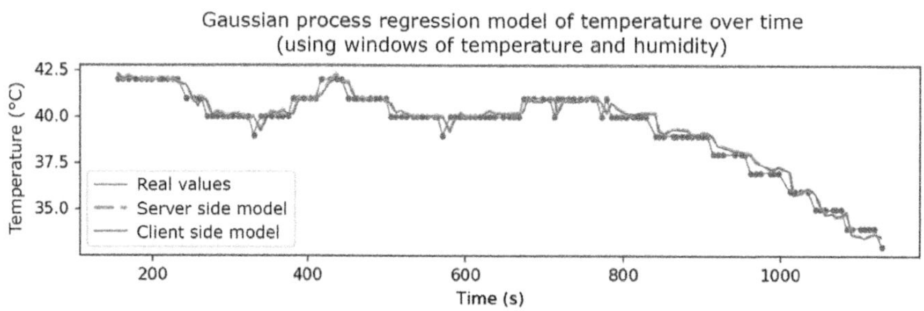

FIG. 7.10

GPR model prediction in server and client sides as well as the real value (value to be predicted). In this figure the model was trained with the third dataset (Table 7.3).

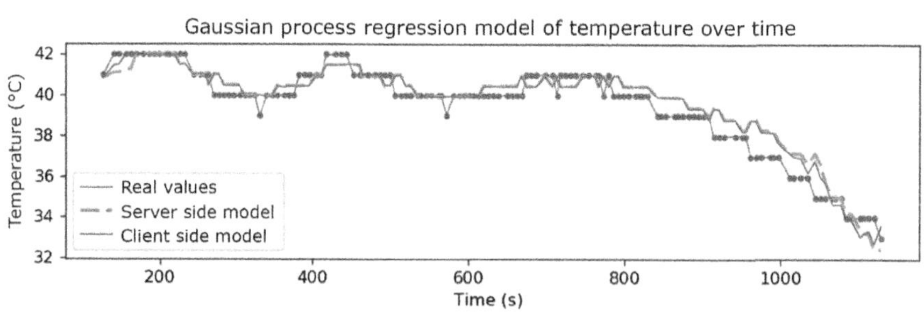

FIG. 7.11

GPR model prediction in server and client sides as well as the real value (value to be predicted). In this figure the model was trained with the first dataset (Table 7.1).

7.7 Conclusions and future work

This chapter proposes an edge-centric predictive methodology that uses online regression model with caching. This method updates the base node models only when significant changes in the edge occur, while sending only the model's parameters instead of raw data. This can downgrade the communication overhead between end nodes and base nodes. After comparative assessment of the models, this chapter presents two highly efficient modeling techniques to be used with the proposed methodology. Our future agenda involves the extension of this methodology to include optimal stopping point theory, in order to search the positive implications of such analysis in the proposed methodology. Another experiment is to discover how the accuracy/communication overhead trade-off per model changes, while the threshold β adjusts its value by a given step. A third experiment is what happens to the accuracy/communication overhead trade-off per model in various levels of packet-loss (e.g., packet-loss = [0.15, 0.30, …,0.75, 0.9]). Lastly, and most anticipated, is to develop the methodology proposed in this literature directly in a MCU. The MCU will read values from a sensor and it will communicate in real time with its base node. From this experiment we are going to be able to accurately measure the energy costs that a MCU is paying to communicate with its base node, and how the proposed methodology will benefit an end node to consume less energy over time.

Acknowledgments

This research has been supported in part by the Huawei Innovation Research Program (Grant No. 300952); by the UK Engineering and Physical Sciences Research Council (EPSRC) projects EP/N033957/1 and EP/P004024/1; by the European Cooperation in Science and Technology (COST) Action CA 15127: RECODIS—Resilient communication and services; and by the EU H2020 GNFUV Project RAWFIE-OC2-EXP-SCI (Grant No. 645220) under the ECFIRE+ initiative.

References

[1] T. Stack, Internet of things (IoT) data continues to explode exponentially. Who is using that data and how?, https://blogs.cisco.com/datacenter/internet-of-things-iot-data-continues-to-explode-exponentially-who-is-using-that-data-and-how, 2018.

[2] B.B. Mishra, S. Dehuri, B. Panigrahi, A. Nayak, B. Mishra, H. Das, Computational Intelligence in Sensor Networks, Springer-Verlag, Berlin/Heidelberg, 2019.

[3] N. Harth, C. Anagnostopoulos, Edge-centric efficient regression analytics, in: IEEE International Conference on Edge Computing, 2018.

[4] K. Panagidi, I. Galanis, C. Anagnostopoulos, S. Hadjiefthymiades, Time-optimized contextual information flow on unmanned vehicles, in: IEEE 14th International Conference on Wireless and Mobile Computing Networking and Communications (WiMob 2018), 2018.

[5] S. Pasteris, S. Wang, C. Makaya, K. Chan, M. Herbster, Data distribution and scheduling for distributed analytics tasks, in: 2017 IEEE SmartWorld, Ubiquitous Intelligence

Computing, Advanced Trusted Computed, Scalable Computing Communications, Cloud Big Data Computing, Internet of People and Smart City Innovation (SmartWorld/SCALCOM/UIC/ATC/CBDCom/IOP/SCI), 2017, pp. 1–6.

[6] K. Kolomvatsos, C. Anagnostopoulos, In-network decision making intelligence for task allocation in edge computing, in: 30th International Conference on Tools with Artificial Intelligence (ICTAI 2018), 2018.

[7] T. Kontos, C. Anagnostopoulos, E. Zervas, S. Hadjiefthymiades, Adaptive epidemic dissemination as a finite-horizon optimal stopping problem, Wirel. Netw. (2018).

[8] N. Harth, C. Anagnostopoulos, D. Pezaros, Predictive intelligence to the edge: impact on edge analytics, Evol. Syst. (2018) 95–118.

[9] M. Ebden, Gaussian processes for regression: a quick introduction, arXiv, 2008. 1505.02965.

[10] C. Rasmussen, C. Williams, Gaussian processes for machine learning, http:/www.gaussianprocess.org/gpml/chapters/RW.pdf?fbclid=IwAR23z0cepMfvjZ_tIENqJzVYVeOfltaRe6KZ1zBTgHuR1HwgkQRRJit2eYA.

CHAPTER

An effective blockchain-based, decentralized application for smart building system management

8

Quanqing Xu*, Zhaozheng He†, Zengxiang Li*, Mingzbhong Xiao‡, Rick Siow Mong Goh*, Yongjun Li§

**Institute of High Performance Computing, A*STAR, Fusionopolis, Singapore †School of Electrical & Electronic Engineering, Nanyang Technological University, Jurong West, Singapore ‡College of Information Science and Technology, Beijing Normal University, Beijing, China §School of Computer, Northwestern Polytechnical University, Xi'an, China*

8.1 Introduction

Blockchain technology maintains a distributed database of records among participating nodes in a distributed network. Transactions are used to change the states in the network via a consensus protocol. The blockchain technology is a combination of many advanced technologies from different areas, such as cryptography, distributed consensus, hashing, networking, and so on. All of these allow the blockchain technology to equip the system with properties like distributed fault tolerance, history immutability, transaction automation, privacy safeguarding, and resiliency to attack. The most significant feature is that the participating nodes attend the maintenance of the distributed database without a centralized, trusted third party. Due to these types of properties, many companies and organizations in different areas have been trying to build new DApps using blockchain. They are trying to apply the great potential of blockchain in cryptocurrency, assets management, financial services, resource sharing, IoT, and many industry domains. The blockchain technology is able to provide excellent distributed trust [1], which is a key enabler for IoT to deliver intended results. Cutting-edge IoT solutions enable us to make better data-driven decisions for smart buildings with blockchain technology.

The Internet is advancing into a new technology era based on blockchain, IoT, and smart systems. It is expected that these technologies will interact and converge in the coming years. The integration of IoT and blockchain will be the most prominent in this evolution since this integration might provide a new way for automated

Real-Time Data Analytics for Large Scale Sensor Data. https://doi.org/10.1016/B978-0-12-818014-3.00008-5

service systems and business models. IoT is the combination of physical devices and information systems to enable machine-to-machine communication and cooperation. Due to its efficiency and potential in the market, the IoT network is growing rapidly and becoming increasingly complex with increasing numbers of well-connected devices. The IoT infrastructure could be used to connect a vast range of devices ranging from autonomous vehicles to sensors monitoring industrial processes to smart buildings and smart cities [2]. There will be 50 to 100 billion devices connected to the Internet by 2020, as predicted by the European Commission [3]. It is the most challenging process to develop an IoT infrastructure that can store and maintain such a large amount of data streaming, even up to EB-scale data [4], with sustainable security capabilities.

There is a huge potential market for managing these devices in a systematic and effective manner. By 2025 the potential market for home automation is estimated to grow to $44 billion, according to Moreno et al. [5]. Therefore smart building that allows building appliance automation is one of the important aspects of IoT in the future. Thus it is necessary to design some efficient and optimal fundamental networking systems. The networking system should allow the devices connected to it to communicate with each other efficiently and perform correct automation actions depending on different conditions. Nowadays most of the IoT foundation systems are cloud-based [6] and they have the problem of high centralization, which is vulnerable to attack. Blockchain can be built as a transaction platform with features like decentralization and high security [7]. Smart contracts could also be implemented in the blockchain to set the rules for automation processes in the real world. Some believe that blockchain is the foundation behind the IoT. Some research has been done to combine smart building and blockchain technology to exploit the advantages of both IoT and blockchain. Sapphire [8] utilizes IoT device computing power to run smart contracts, which perform computations on the data coming from a smart IoT environment.

In this chapter, a system prototype named Uranus was built to show a possible combination of IoT and blockchain. This system prototype simulates smart building applications with smart contracts and data storage. RPis are used as edge-computing servers to acquire data from the DHT11 temperature and humidity sensors every 15 s. The captured data is then display in the Blynk app,[1] which is a mobile digital dashboard for IoT projects that allows the user to monitor and interact with the smart contract. At the same time, the collected sensor data is passed to the smart contract to be executed on Ethereum. This system prototype proposes a new way for data storing and reading of smart building applications. RPi 3 sends and reads the data to and from the blockchain by running a JavaScript program in the *Node.js* runtime environment. Users can also set the threshold values for both temperature and humidity, which are stored on the smart contract. Whenever these values change, the JavaScript program running on the RPi compares the real-time temperature and humidity values with the

[1]https://www.blynk.cc/.

corresponding threshold values set by the users on the smart contract. When the real-time temperature or humidity value exceeds the corresponding threshold value, the red LEDs or green LEDs turn on as an indication of warning. This is a demonstration system as a smart building automation application that turns on the air conditioner or dehumidifier in a room when the corresponding temperature or humidity exceeds the corresponding threshold.

This chapter is organized as follows. Section 8.2 describes the background and Section 8.3 describes overall design of our proposed system. Section 8.4 presents the system implementation and Section 8.5 reports the performance evaluation results. Section 8.6 reviews the related work and Section 8.7 concludes the chapter and presents some future work.

8.2 Background

8.2.1 Ethereum

Ethereum[2] was created as a platform intended to allow developers to easily build decentralized applications (DApps) with blockchain technology. Ethereum adds in a built-in Turing-complete programming language, Solidity [5], for the developers to set the rules of DApps by coding them down as a smart contract. A smart contract is one type of account existing in Ethereum that stores the code in the blockchain and executes it automatically when other accounts call it. A DApp provides fluent transactions with the smart contracts being executed automatically under the circumstances specified in advance. Ethereum also allows the DApps to run on three types of networks: public blockchains, consortium blockchains, and private blockchains, depending on the purposes of the applications.

At the time of writing, Ethereum is still using Proof of Work (PoW) as the consensus protocol. For PoW, the miners in the network compete by repeating the same process, which is hashing the unique header metadata with changing "nonce value." This process returns a hash value, which is compared with the current target in the Ethereum blockchain. Only when the current target is met, will the miner stop the process and broadcast the result to the whole network. All the nodes validate the block and add it to their own copy of the ledger. The miner who just mined the block will receive ethers as a reward. The rest of the miners will stop and repeat the process for the next block. The Ethereum community is trying to shift the consensus protocol to Proof of Stake (PoS). A smart contract is the code created and stored in a specific address by the developers of the DApp. Since smart contracts have great capability, developers have been trying to use them to automate the exchange process of money, property, shares, or any types of value in a transparent, conflict-free way without an

[2]https://www.ethereum.org/.

intermediary. Many believed smart contracts could help revolutionize finance services, credit and ownership management, resource sharing, investment management, IoT, and other areas.

8.2.2 Internet of things

The IoT is changing much about the world we live in. Both IoT and Industry 4.0 are disruptive innovations to our society. They are believed to be able to increase the efficiency of not only factories but of whole societies. The areas that could be improved by the IoT are vast, including autonomous driving, grocery shopping, energy distribution, home appliance control, and so on. The IoT enables all the physical tools and devices embedded with sophisticated sensors [9], multifunctional chips, and network connectivity to communicate and exchange data within the same network. The valuable data transmitted between all these devices allow us to execute different tasks automatically depending on the situation. This also provides us a path to understand how these things work and cooperate.

Some examples of the potential of IoT include improving industry production efficiency via automatic machine-to-machine cooperation, giving city residents real-time updates on where to park, and monitoring our health. The common IoT platform [10] brings the diverse information together and provides the common languages for the devices and applications to communicate with each other. It would certainly improve process efficiency since more time and money would be saved in manufacturing, business, and our daily lives. A private or consortium blockchain network can help us to create a tamper-resistant record. This opens new ways of automating business processes among IoT devices without setting up a costly centralized IT infrastructure.

8.3 Overall design structure

In this section, we show the details of the hardware and libraries used in this system, and we discuss the justifications for choosing the specific hardware and JavaScript libraries. The JavaScript code is written to send the sensor data to the Blynk app, call the function on the smart contract to store the sensor data on chain, and control the LEDs depending on the temperature and humidity data stored on the smart contract. The JavaScript files are run on the *Node.js* environment on the RPi 3 Model B.

8.3.1 Overall workflow

The overall design workflow of the system Uranus is shown in Fig. 8.1. Four major components are integrated to build this smart building system. They are DHT11 temperature and humidity sensors, the Blynk app, RPi, and the Ethereum private blockchain. The DHT11 sensors measure the room temperature and humidity and send the data to the RPi every 15 s. Ethereum is the platform used to build the private blockchain and smart contract. The private Ethereum blockchain is maintained by the miners. The smart contract written for this system prototype is to store the current sensor and threshold values submitted fro.m the RPi. Secure multiparty computation [11] and Non-Interactive Zero-Knowledge (NIZK) proofs, for example, Accumulator Proof of Knowledge (AccPoK) and Serial Number Signature of Knowledge (SNSoK) [12], are used to protect privacy and anonymize transactions in Uranus.

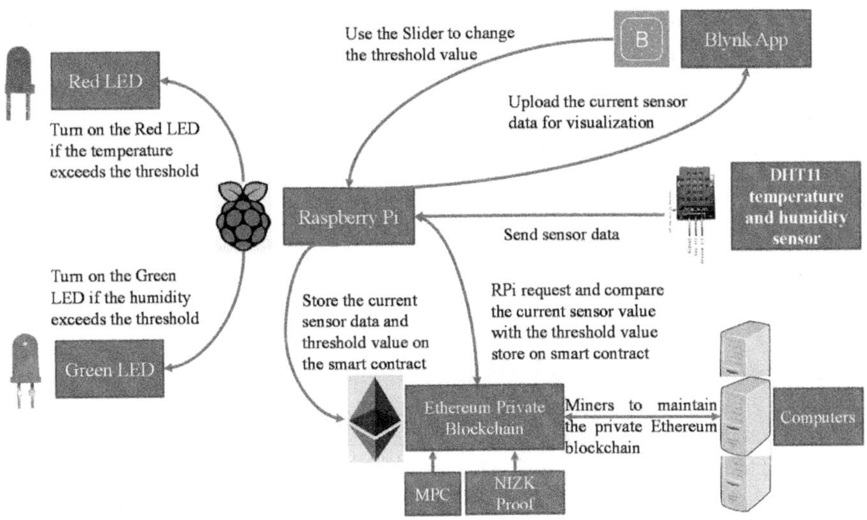

FIG. 8.1

Overall workflow of *Uranus.*

8.3.2 Raspberry Pi

RPi is a low-priced, single-board computer similar in size to a credit card developed by the Raspberry Pi Foundation in the United Kingdom. Originally it was designed to promote the teaching of basic computer science in schools and in third-world countries. Developers and hobbyists realize that its flexibilities and cost-effectiveness make it ideal for creating projects. It is starting to be used in fields such as robotics and teaching, among others. According to the Raspberry Pi Foundation, 19 million RPis were sold by March 2018, making it the best-selling British computer.

8.3.3 DHT11

DHT11 is a temperature and humidity sensor complex with a calibrated digital signal reading. There is only one wire for the data signal, which makes it very easy to set up. DHT11 is popular for use in remote weather stations, soil monitors, and home automation systems. The DHT11 is a composite of a resistive-type humidity measurement component and a surface-mounted NTC thermistor. A high-performance 8-bit microcontroller on the back of the module converts the resistance measurements from the thermistor, humidity measurements into digital temperatures (in °C) and relative humidity measurements. It provides advanced properties such as excellent quality, cost-effectiveness, fast response, and antiinterference ability.

There are two types of DHT11 on the market: the four-pin version and the three-pin version. We chose the three-pin version for our system because there is a surface-mounted 10K Ohm pull-up resistor for the signal line on it. Users can mount the DHT11 on the printed circuit board (PCB) board easily. The *node-dht-sensor* is a JavaScript library that supports querying temperature and relative humidity from compatible DHT sensors. This module depends on the BCM2835 library that must be installed before the *node-dht-sensor* library can be used. In order to initialize the sensor, the sensor type and the General-Purpose Input/Output (GPIO) pin where the sensor is connected are to be specified at the beginning. After the initialization, the function.*read*() can be called to obtain the latest temperature and humidity readouts from the sensors. The *onoff* is a JavaScript library for a developer to access the GPIO and detect interrupt with *Node.js* on Linux boards like the RPi. The GPIO pin of RPi can be controlled by the *onoff* module, either sending output or reading input.

8.3.4 Blynk

Blynk is a platform that provides iOS and Android apps for users and developers to control ESP8266, Arduino, RPi, SparkFun, and many other microcontrollers in the market. Developers are allowed to drag and drop the widgets onto the dashboard to build a graphical user interface (GUI) for their projects. These widgets can be used to control the hardware remotely as well as to visualize and display sensor data. It is an end-to-end solution for the IoT. It reduces the time and resources needed to build the applications for connected products and services.

The Blynk platform enables organizations to move smoothly from prototype to production in short iterations. It also allows developers to collect feedback and refine the product at every development stage. Blynk is secure, scalable, lightweight, and fast with the capability of managing billions of requests to and from edge devices. The Blynk library is a JavaScript library used to communicate between the Blynk server and the hardware, which in this system is RPi. It is used to receive temperature and humidity data from the sensors and displays the data on the Blynk app.

8.3.5 Ethereum components

Ethereum Virtual Machine (EVM) serves as the runtime environment for smart contracts based on Ethereum. It protects the smart contracts from any potential interference by ensuring they do not have access to each other's state. That means the EVM is completely isolated from the rest of the main net; every Ethereum node runs only their own EVM implementation with the same instruction set. All of these properties make EVM an ideal environment for smart contract development testing without affecting operations in the main blockchain. Ethereum clients refer the command lines tool or GUI software that provides interaction capabilities with the Ethereum network. This means the user could use the Ethereum clients to send transactions and start mining. There are three official command line tools for Ethereum: C++, GoLang, and Python. They all provide the same functionalities; the only difference is that they are implemented in different languages. Mist is the only official graphical client that allows the user to interact with the Ethereum blockchain without typing the instructions. For this project, we used the GoLang client Geth since it is the most popular client amongst developers.

In Ethereum, there are two types of accounts: (1) Externally Owned Accounts (EOAs), which use the normal Ethereum address controlled by the user's key for the user to send transactions and (2) Contract Accounts, which is the address used to store the code of the smart contract. The smart contract at that particular address could be triggered by the transaction from EOAs or messages from other contract accounts. Mining is a process for the miners to validate a block created in the network. It is an incentive-driven model since the miners would receive the appropriate rewards according to the work they made. The security of the network improves as the number of miners increase. Web3 is a JavaScript library that allows us to interact with a local Ethereum node, using an HTTP or IPC connection. It is available on *npm* as a node module. In our system, it is used to call the functions on the smart contract from the RPi.

8.3.6 Put them all together

RPi is a single-board computer that can simply perform any task just like a regular computer. We use RPi 3 model B 2015 in this system. Two JavaScript files are used for RPi to interact with DHT11, the Blynk app, and the private Ethereum blockchain. RPi collects the sensor data from DHT11 and visualizes them in the Blynk app. When new sensor data is received, RPi also uploads the sensor data to the smart contract. It compares the current temperature and humidity values with the respective threshold values stored on the smart contract. If the current temperature value exceeds the temperature threshold, the red LED is turned on. Similar procedures apply to the humidity value, but using a green LED instead. RPi interacts with the smart contract and changes the corresponding threshold value if the users change it on the Blynk app via the slider widget. Ethereum, as a blockchain platform, has three advantages: (1) it is tamper-resistant, (2) it has no single point of failure, and (3) it is trusted and auditable. Therefore it is ideal for building a DApp for smart building system management.

8.4 System implementation

8.4.1 Set up the Raspberry Pi

This step will transform each RPi 3 into an Ethereum node but not a mining node. Use the following steps to set up RPi:

(1) Insert the SD card into the MacBook and format it as ExFat using the disk (take MacOS for example).

(2) Download the NOOBS (New Out Of the Box Software), which is one of the official operating systems for RPi, from the official website. Extract all the files from the ZIP file and copy all the files onto the SD card.
(3) Insert the SD card into the RPi and connect the RPi touchscreen to the RPi via the HDMI port. When the RPi is powered on, the NOOBS OS should start automatically and we can start to use it with our mouse and keyboard connected.
(4) Reset the default password, keyboard layout, and time zone accordingly. Enable SSH such that the computer can access the RPi securely later. Change the hostname to $node_n$ for it to identify the RPi within the network easily. These can be done by typing the command into the command line tools of RPi. We can use the command *sudo raspi-config* for a configuration to make the changes.
(5) Connect the RPi to Wi-Fi, and check the IP address of the RPi by using the following command: *ifconfig*.
(6) Download the appropriate version of Geth for the RPi. Since the model of the CPU of the RPi is ARMv7, the ARMv7 version of Geth should be downloaded. Extract the bundles of files from the ZIP and copy the Geth application to */usr/local/bin* so that the built-in command line tools of RPi can use the Geth command.

8.4.2 Installation of Ethereum

We installed Ethereum nodes on the computers and RPi devices. Miners are needed in the private blockchain to validate the blocks of transactions. Miners are also needed to collect the token rewarded from the system such that the other nodes could have ether to pay for the gas required for the transactions. The mining process is too intense for the RPi so that at least two miners are needed to be deployed on the computers. Each of the nodes on the computers utilizes a specific folder to store the database of the private blockchain without affecting the public blockchain data folder. Both nodes initialize the private blockchain based on the same genesis file with the same network ID to make sure they are in the same private blockchain. Of course, the port numbers of the nodes must be different to differentiate them on the same computer.

8.4.2.1 Genesis JSON file

As shown in Fig. 8.2, we created the Genesis JSON file, which is used to initialize the private blockchain and define the important parameters of the blockchain, including block difficulty, gas limit, network ID, and so on. As shown in Fig. 8.2, the *genesis.json* file is created under the folder *~/SmartBuilding*.

As stated in the genesis file, the network ID is 65. The difficulty is set to 0×200, which is relatively low so that the transactions will be processed faster in the private

```
{
  "config": {
    "chainId": 65,
    "homesteadBlock": 0,
    "eip155Block": 0,
    "eip158Block": 0
  }
  "mixhash": "0x0000000000000000000000000000000000000000000000000000000000000000",
  "nonce": "0x0000000000000065",
  "difficulty": "0x200",
  "gasLimit": "0xffffffff",
  "alloc": {},
  "extraData": "",
  "timestamp": "0x00",
  "coinbase": "0x0000000000000000000000000000000000000000",
  "parentHash": "0x0000000000000000000000000000000000000000000000000000000000000000",
}
```

FIG. 8.2

Genesis block JSON data.

blockchain. The gas limit is set to maximum to prevent the transactions from failing by exceeding the gas limit. We initialized the miners with the genesis file and generated an account for each node. A subfolder was created in the data directory after the initialization. The *geth* folder stores a copy of the database of the private blockchain for each node. The *keystore* folder stores the account related to the corresponding node.

8.4.2.2 Ethereum miner

For a miner, a *password.sec* file containing the password of the account is created under *~/SmartBuilding/miner*. This file is used to start the miner easily. We create a runnable script called *startminer.sh* (as shown in Fig. 8.3) in *~/SmartBuilding/miner* that stores the start command for the miner so that we do not need to retype thc chuck of command every time.

```
#!/bin/bash
geth --networkid 65 --identity "miner1" --mine --nodiscover --rpc --rpcport "8042" -
-port "30303" --unlock 0 --datadir "~/SmartBuilding/miner" --ipcpath
"~/Library/Ethereum/geth.ipc" --password "~/SmartBuilding/miner/password.sec"
```

FIG. 8.3

Starting a miner (startminer.sh).

From the runnable script, the identity of this node is set to *miner1* and the network ID is 65, which is the same as the genesis file discussed previously. The data directory to save the copy of database for the *miner1* is *~/SmartBuilding/miner*. The *rpcport* is set to 8042, which enables the HTTP-RPC server. The port number is set to 30303 as the network listening port number for the nodes to connect each other and spread the transactions and blocks. The discovery mechanism is disabled because the nodes are going to be connected using the static connection. The node is set to a miner with the option *–mine*. The option *–unlock 0* sets the default account as the 0th account. The *password.sec* file is stated after the *–password* such that the account will be unlocked with the password saved in the file. The *ipcpath* defines where to store the *geth.ipc* file, which is for IPC socket/pipe. It allows Mist to detect the private network and connect to it instead of the public main net.

8.4.2.3 Ethereum clients in Raspberry Pi

We proceeded to set up an Ethereum node in the RPi. First, we transfered the *genesis.json* file to the RPi. Second, we initialized the node and created two accounts for it. Then we created the runnable script *startnode.sh* (as shown in Fig. 8.4) for the node under the directory *~/SmartBuilding/node*.

The procedure for setting up a new node in another RPi is similar. The only difference is the identity should be changed to $node_n$.

```
#!/bin/bash
geth --networkid 65 --identity "node1" --fast --nodiscover --rpc --rpcport "8042" --
port "30303" --unlock 0 --datadir "/home/pi/SmartBuilding/node" --ipcpath
"/home/pi/.ethereum/geth.ipc" --password "/home/pi/SmartBuilding/node/password.sec"
```

FIG. 8.4

Starting a fast node (startnode.sh).

8.4.3 Pair all the nodes in the private blockchain

Blockchain is a peer-to-peer network, which means the miners need to be aware of each other for the transaction propagation. In this system, the configuration is made to specify the identity and the location of each node's peers. We identified the IP address of the computer on which the miners are running using the following command: *ipconfig getifaddr en0*. We obtained individual node information on the computer in the JavaScript console using the command *admin.nodeInfo.enode*. We retrieved the node information from RPi with the same command: *admin.nodeInfo.enode*. We created the file *static-node.json* under the data directory of each node. The file *"static-node.json"* defines the permanent static nodes in the private blockchain. When the node starts, Geth processes the file automatically and connects the nodes together. We checked the synchronization process of all the nodes from a miner.

8.4.4 Create and deploy the smart contract RPi_DHT11_2

Our smart contract stores the temperature and humidity of the two sensors. Events are used to log the historical data of the temperature and humidity. When there is an event change, the RPi compares the current temperature and humidity with the user-set threshold values. Whenever the current value exceeds the threshold, the corresponding LED lights up as a warning.

We tested the smart contract on the official online Ethereum IDE Remix[3] and deployed it using the Truffle[4] development framework. We constructed and tested the smart contract on Remix and made sure it worked properly. Fig. 8.5 shows the variables and functions defined in the smart contract. The basic logic workflow is depicted by the arrows.

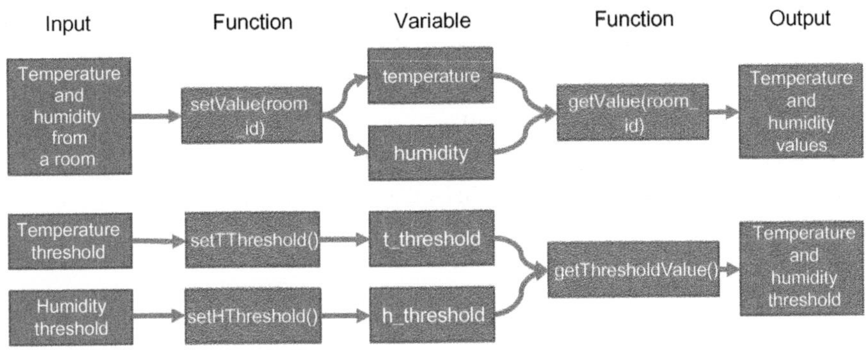

FIG. 8.5

Structure of the smart contract.

There are five functions in this contract:

(1) *setValue()*: update the temperature and humidity value of a room.
(2) *setTThresold()*: set the temperature threshold.
(3) *setHThresold()*: set the humidity threshold.
(4) *getValue()*: retrieve the sensor data store on the smart contract.
(5) *getThresoldValue()*: retrieve the threshold value set by the user.

We installed Truffle, created a project, and created a smart contract file named *RPi_DHT11_2.sol*. We created a file called 2_deploycontracts.js as shown in Fig. 8.6. We started the miners to ensure that the smart contract would be mined properly after it was deployed to the private blockchain. After that, we compiled and deployed the smart contract.

```
var RPi_DHT11_2 = artifacts.require("./RPi_DHT11_2.sol");

module.exports = function(deployer) {
deployer.deploy(RPi_DHT11_2);
};
```

FIG. 8.6

Deploying a smart contract.

[3]https://github.com/ethereum/remix.
[4]https://github.com/trufflesuite/truffle.

We interacted with the smart contract on Mist. The address and ABI of the contract must be known before we can use Mist to interact with the contract. We started Mist and selected Watch Contract. We put in the contract address and ABI retrieved as shown in Figs. 8.7 and 8.8.

FIG. 8.7

Adding a smart contract.

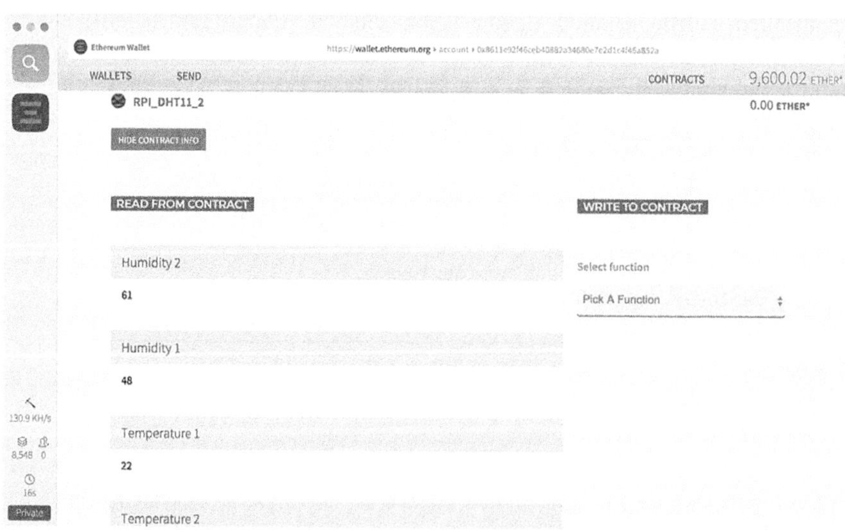

FIG. 8.8

Smart contract interface.

8.4.5 Connect the Blynk with RPi and DHT11

Blynk is a universal remote-control system, and it is used to collect the data from the DHT11 temperature and humidity sensors via the RPi. Our prototype was made to simulate the smart building applications that collect the temperature and humidity conditions of separate rooms and control the corresponding LEDs according to the situation. The JavaScript library *blynk-library* is used to send the sensor data to the Blynk app. We installed the *Node.js* in the RPi and installed the required JavaScript libraries: *onoff* and *blynk-library*.

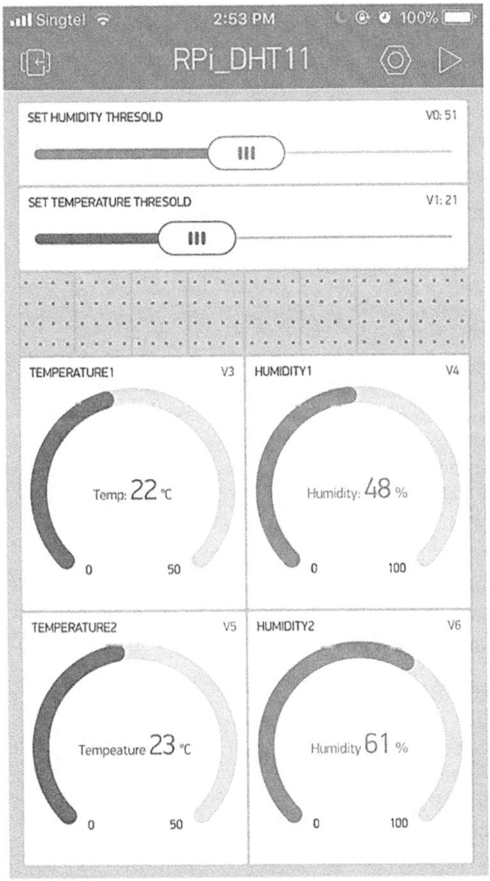

FIG. 8.9

The Blynk App of the project.

We created a new project in the Blynk app and added four gauge displays and two sliders to the project as shown in Fig. 8.9. The first slider binds to *V0* and is used to set the humidity threshold and the second slider binds to *V1* and is used to set the temperature threshold, as shown in Fig. 8.10. The two gauge displays are called

Temperature1 and *Humidity1* (as shown in Fig. 8.9) and are bound to *V3* and *V4* to collect the temperature and humidity values from the first DHT11 sensor via the first RPi. As shown in Fig. 8.9, the two gauge displays called *Temperature2* and *Humidity2* are bound to *V5* and *V6* to collect the temperature and humidity values from the second DHT11 sensor via the second RPi.

```
var blynkLib = require('blynk-library');

//Setup Blynk
var AUTH = 'a36d16d5ce6242668ccc09e4f09fe2b2';
var blynk = new blynkLib.Blynk(AUTH);

//V0 is the value of Slider from the Blynk App
var v0 = new blynk.VirtualPin(0);
//V1 is the value of Slider from the Blynk App
var v1 = new blynk.VirtualPin(1);

var sensorLib = require('node-dht-sensor');
//Setup sensor, exit if failed to initialize sensor
...

//Update sensor value every 15 seconds
setInterval(function() {
    var readout = sensorLib.read();
    blynk.virtualWrite(5, readout.temperature.toFixed(1));
    var _temperature = parseInt(readout.temperature.toFixed(1));
    console.log('Temperature: ', _temperature);
    blynk.virtualWrite(6, readout.humidity.toFixed(1));
    var _humidity = parseInt(readout.humidity.toFixed(1));
    console.log('Humidity: ', _humidity);
    contract.setValue2(_temperature, _humidity);
}, 15000);

v1.on('write',function(param){
    console.log('Temperature Thresold: ', param[0]);
    contract.setTThresold(param[0]);
});

v0.on('write',function(param){
    console.log('Humidity Thresold: ', param[0]);
    contract.setHThresold(param[0]);
});
```

FIG. 8.10

The first slider binds to V0, and the second slider binds to V1.

We set up the physical connections for the RaPi, DHT11, and LEDs. A RPi reads as shown in Table 8.1.

Table 8.1 Connection of GPIO for a Raspberry Pi

Component	GPIO number of RPi
DHT11	The data pin is connected to GPIO4. The "+" pin is connected to 3.3V, and the "−" pin is connected to ground
Anode of Green LED	GPIO14
Anode of Red LED	GPIO15
Cathode of LEDs	Ground

We generated the corresponding JavaScript file to collect sensor data from DHT11, display the data on the Blynk app, and control the red and green LEDs. All the JavaScript files that use the Blynk app must include the *blynk-library* and define the authentication token corresponding to the Blynk project. The *node-dht-sensor* library is used to read the sensor data. Definition of the type of sensor and the GPIO pin connected to the data pin of DHT11 is needed. The sensor data is uploaded to the Blynk app every 15 s since the average block time in Ethereum is 15 s. The *onoff* JavaScript library is used to control the LEDs. GPIO pins connected to the LEDs are needed in the initialization. The new sensor data is compared to the threshold value and actions are taken according to the result. When the temperature exceeds the threshold, the red LED is turned on. When the humidity exceeds the threshold, the green LED is turned on. The flowchart of the coding implementation is shown in Fig. 8.11.

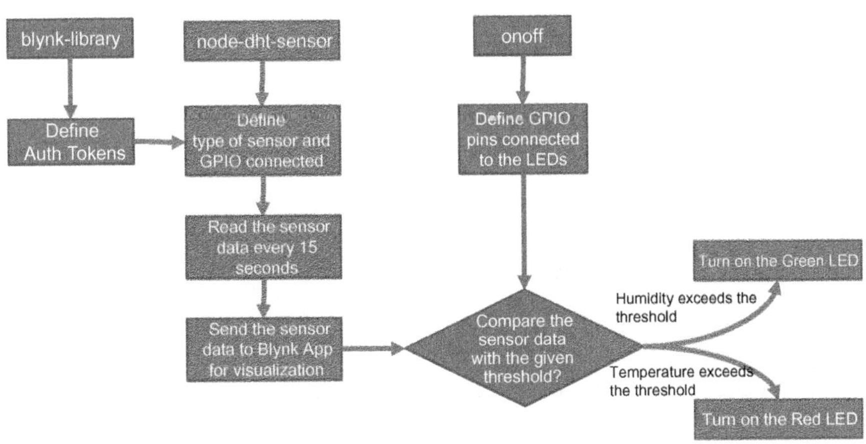

FIG. 8.11

Flowchart for connecting the Blynk with RPi and DHT11.

8.4.6 Build JavaScript applications on RPi to interact with smart contract

In this step, the library *web3* is used to access the smart contract functions. We set up the connection from the JavaScript file to the smart contract. All the files that interact with the smart contract need to initialize with the address and application binary interface (ABI) of the smart contract. The sensor data is uploaded to the smart contract every 15 s and the threshold value is updated whenever the users change the slider in the Blynk app. The value of temperature and humidity stored on the blockchain is compared to the corresponding threshold whenever they are updated. The function *showStatus()* is called to turn on or off the appropriate LEDs, as shown in Fig. 8.12.

```
//Power a LED according the value of the token
function showStatus() {
    //Display the LED according the value of the token
    if (temperature > t_thresold) {
        //Red: Temperature exceeds the thresold, turn on the aircon
        redLed.writeSync(1);
    }
    else{
        redLed.writeSync(0);
    }
    if (humidity > h_thresold){
        //Green: Humidity exceeds the thresold, turn on dehumidifier
        greenLed.writeSync(1);
    }
    else{
        greenLed.writeSync(0);
    }
}
```

FIG. 8.12

Status function.

The flowchart of the full implementation of the JavaScript is shown in Fig. 8.13.

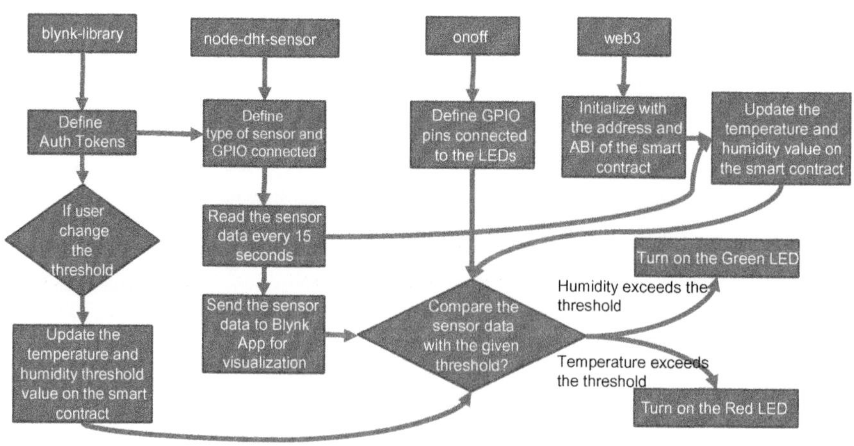

FIG. 8.13

Flowchart of full implementation of the JavaScript file.

8.4.7 Run the PoC system

A private Ethereum blockchain is set up for this system. There are miners from multiple computers and nodes from the RPi 3 devices. The system can collect the temperature and humidity data from the DHT11 sensors via the RPi devices. The collected data is sent and stored on the smart contract named *RPi_DHT11_2*. At the same time, the values are displayed at the corresponding gauge widget at the bottom of the Blynk app as shown in Fig. 8.9. Users can also use the Blynk app to change the temperature and humidity thresholds via the sliders at the top of the app. Whenever the user changes the value of the slider, the RPi calls the corresponding function on the smart contract, either *setTThresold()* or *setHThresold()* (as shown Fig. 8.14) to change the threshold. The RPi checks the values of temperature and humidity whenever they are changed on the smart contract. If the values exceed the threshold set by the user, the RPi turns on the corresponding LED as an indication of warning. The LED is turned off whenever the room temperature or humidity value is lower than the preset threshold.

```
pragma solidity ^0.4.21;

contract RPi_DHT11_2 {
    ...
    event onTThresoldChange(
        uint e_t_thresold
    );
    event onHThresoldChange(
        uint e_h_thresold
    );
    ...
    function setTThresold(uint _t_thresold) public {
        t_thresold = _t_thresold;
        onTThresoldChange(_t_thresold);
    }
    function setHThresold(uint _h_thresold) public {
        h_thresold = _h_thresold;
        onHThresoldChange(_h_thresold);
    }
    ...
}
```

FIG. 8.14

setTThresold() and setHThresold().

The PoC system can be used to simulate a smart building application. The DHT11 sensors connected to the separate RPi devices can be placed in multiple separate rooms in a building such that the temperature and humidity of each room can be monitored. The values are stored on the smart contract in the private Ethereum blockchain. If the room temperature or humidity exceeds the threshold set by the user, the corresponding LEDs will be turned on as warnings. Of course, in the real smart building application environment the LEDs should be changed to air conditioner or dehumidifier. So when the temperature exceeds the temperature threshold, the air conditioner will be turned on automatically and vice versa. If the room humidity exceeds the humidity threshold, the dehumidifier will be turn on automatically and vice versa.

8.5 A proof-of-concept case study

In order to demonstrate the effectiveness and feasibility of the proposed blockchain-based smart building system Uranus, we build a real-world testbed made of low-cost and easy-to-get hardware. Firstly, we provide the details of our testbed. Secondly, we present our experimental results.

8.5.1 Testbed setup

Our testbed setup is shown in Fig. 8.15, and our testbed includes DHT11 sensors, RPi devices, air conditioners, dehumidifiers, a private Ethereum network, and the Blynk app in smart building environments. In our testbed, RPi devices are used as edge servers, and they are in charge of receiving requests and processing data. The specifications of this RPi 3B model are shown in Table 8.2.

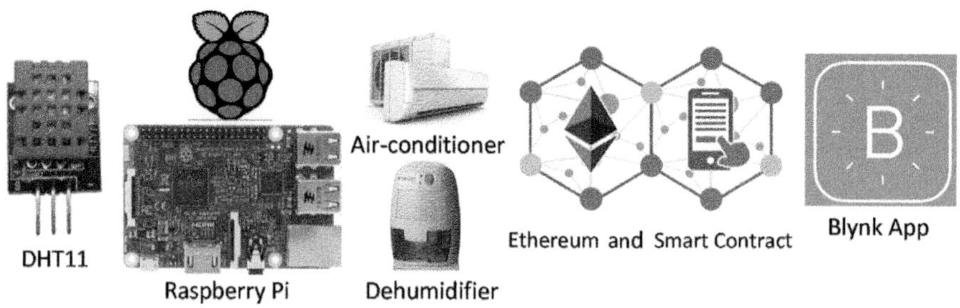

FIG. 8.15

Testbed.

Table 8.2 Raspberry Pi 3 model B specification

Component	Specification
CPU	Quad Core 1.2 GHz Broadcom BCM2837 64 bit
RAM	1 GB
WLAN	BCM43438 on board
Bluetooth	Bluetooth low energy on board
GPIO	40-pin extended
USB	4 USB 2 ports
HDMI	Full-size HDMI
Touch screen	DSI display port for connecting a RPi touchscreen display
Memory	Micro SD port for loading OS and storing data

8.5.2 Performance evaluation

Testing was done on the system for a typical 20 periods. The testing parameters and results for these 20 periods are stated as follows: *T1* and *T2* mean the temperature value of room 1 and 2, *H1* and *H2* are the humidity value of room 1 and room 2, *TT* is the temperature threshold value, and *HT* is the humidity threshold value. GLED refers to the green LED status while RLED refers to the red LED status. The system was built successfully since the sensor data could be stored on the smart contract and displayed on the Blynk app. In this system, a smart building prototype system is set up. RPi is used in this system to collect the sensor data, send it to the Blynk app, and call the appropriate smart contract function at the corresponding situation.

Figs. 8.16 and 8.17 demonstrate room 1's temperature change and its control, and room 2's temperature change and its control, respectively. The LED would be turned on or off in the correct condition. The PoC system shows real-time temperatures of two rooms on the Blynk app with the data collected from two DHT11 sensor devices. The sensor data is uploaded to the Ethereum smart contract RPi-DHT11_2. The smart contract is deployed in the created private blockchain with the miners from the computers and one normal node from each RPi. If the real-time temperature value exceeds the threshold set by the user, the red LEDs will be turned on as warnings to turn on the air conditioners.

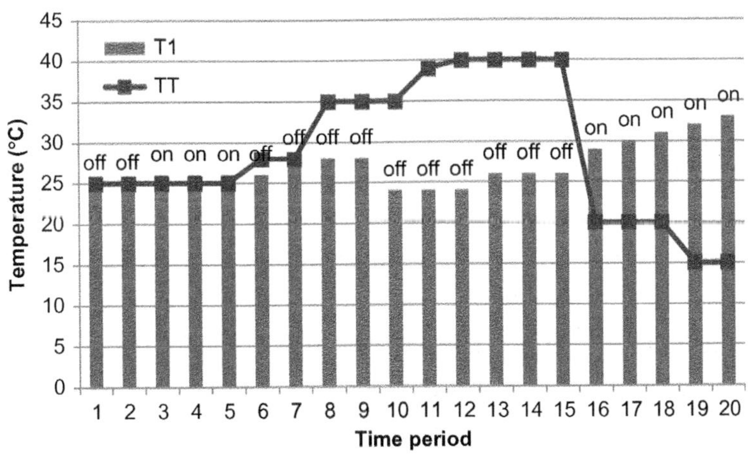

FIG. 8.16

Room 1's temperature change and its control.

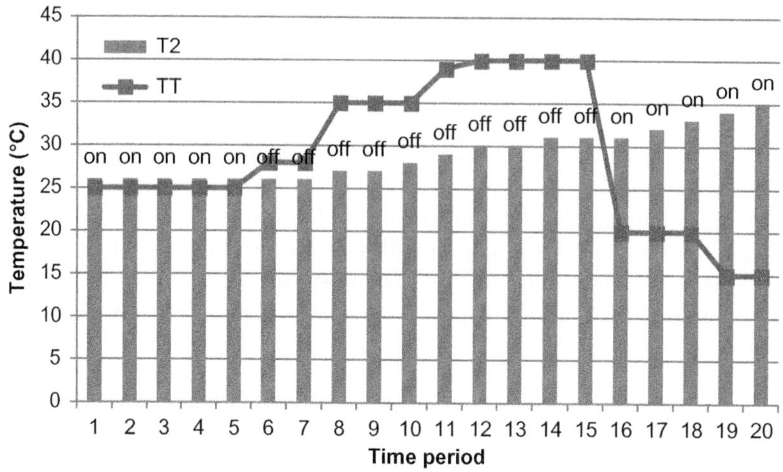

FIG. 8.17

Room 2's temperature change and its control.

Figs. 8.18 and 8.19 show room 1's humidity change and its control, and room 2's humidity change and its control, respectively. After the data are collected from the DHT11 sensor devices, our PoC system demonstrates the real-time humidity values of two rooms on the Blynk app. The collected sensor data is sent to the Ethereum smart contract RPi-DHT11_2. The smart contract is deployed in the created private blockchain with the nonminers from RPi devices and the miners. If the real-time humidity value exceeds the threshold set by the user, the green LEDs will be turned on as warnings to turn on the dehumidifiers.

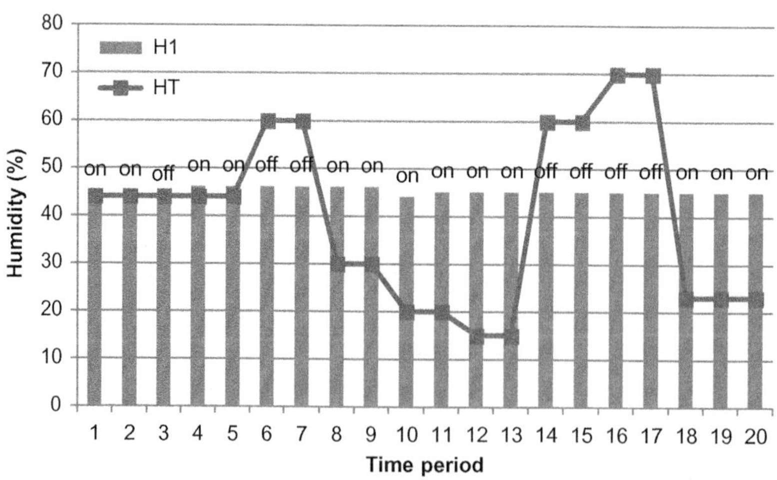

FIG. 8.18

Room 1's humidity change and its control.

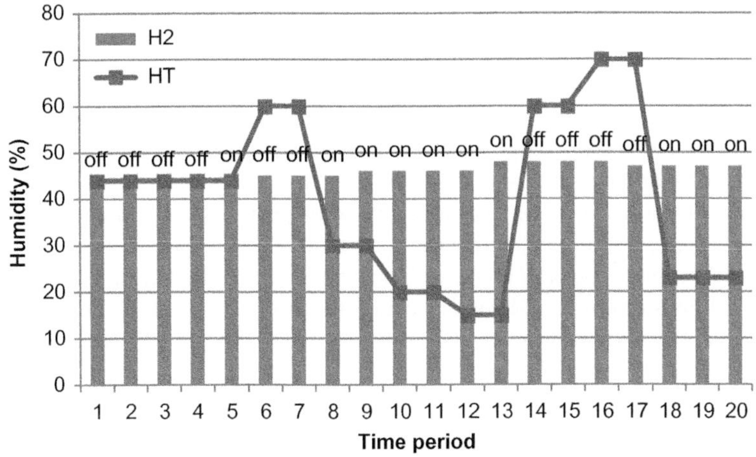

FIG. 8.19

Room 2's humidity change and its control.

Fig. 8.20 shows block-generation time. The block time records the time for the block needed to be added to the blockchain. The average time for block generation is 18.55 s for these 20 periods. The Etheruem block is mined in 15 s on average, which means that the sensor value is uploaded to the smart contract every 15 s. Although 15 s seem to be too long for the approximate real-time data stream, it is still acceptable in this scenario. It is not considered an unacceptable delay if the warning is only shown 15 s on average after the real temperature and humidity of a room exceed the thresholds.

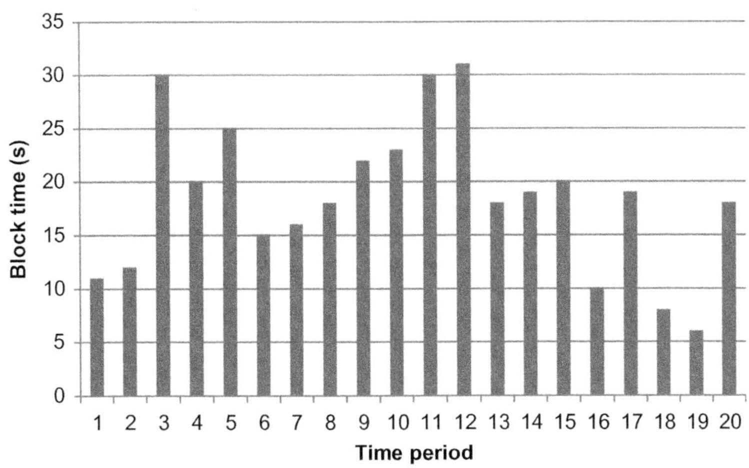

FIG. 8.20

Block-generation time.

8.6 Related work

8.6.1 Blockchain-based IoT

The blockchain is a system or network that maintains a secure distributed database for transaction records. A large number of nodes are organized to create the blockchain to record and authenticate exchange of value. Within the blockchain network, any kinds of value could be exchanged directly peer-to-peer without centralized authorities for recording and validating transactions. Therefore the next generation of IoT systems might be built on top of blockchains with automatic and secure transactions. Billions of connected devices could be tracked by this kind of system with smooth transactions and automatic coordination among devices. The problem of single point of failure can be eliminated as all the participating nodes take part in recording and validating the transactions. The transactions would be more secure with a system more resilient to attack for different kinds of devices to run on. Additionally, the blockchain-based IoT infrastructure provides automatic transactions with fully autonomous device corporations and reduced cost and settlement time [13]. IOTA[5] is claimed to be fast and transaction-free, and it is possibly used to solve the inefficiencies of the blockchain by powering a secure, scalable, and feeless transaction settlement layer.

As blockchain and the IoT are gaining popularity, companies and start-up teams are using them to develop new systems and software. Some of the most well-known development teams aim to solve different kinds of demands in the market. Slock.it[6] is trying to fully automate Airbnb's rental and return process using a smart contract to control door locks remotely. Oaken Innovations[7] is cooperating with the Toyota Research Institute to build a decentralized car leasing application that could provide secure and fast car rental transactions. Filament[8] aims to provide the hardware and software solution for effective machine-to-machine communications, which would enable enterprises and industries to use data to create new efficiencies and cost savings at scale. Xu et al. [8] presented Sapphire, which is a blockchain-based storage system for data analytics in the IoT. Huh et al. [14] proposed using the Ethereum blockchain platform and smart contracts to manage IoT devices that track meters and set policies to turn on and off air conditioners and light bulbs to save energy. Li et al. [15] investigated the convergence of relevant technologies, such as cloud computing, IoT, data analytics, and blockchain, by presenting a prototype of fine-grained blockchain-based transportation insurance.

[5]https://iota.org/.
[6]https://slock.it/usn.html.
[7]https://www.oakeninnovations.com/.
[8]https://filament.com/.

8.6.2 Blockchain-based smart building systems

Lazaroiu et al. proposed a smart district model to provide power grid access to an individual by combining IoT and blockchain [16]. It is a prototype model for users to interact with power grid systems via the blockchain. Any individual with a solar panel could participate in the network to buy or sell energy directly on the blockchain. This is a good example of blockchain-based IoT applications that could be implemented and replicated in the real world. Their paper also defined some important features necessary for a home automation system, which is a good reference for designing a new smart home application. The system should have supervision and control over all the devices in the network. Different types of sensors and actuators should be implemented for the data stream required for the IoT scenario [16]. The system must be intelligent, communicate effectively, and be resilient to malicious attack. A user-friendly interface is necessary for users to interact with this system and to hide the sophisticated algorithm. There should be enough memory space to store all the important data and parameters. The communication among devices should be fluent and effective. It is better to design the system in such a way that it is easily scalable in the future and allows different systems to interoperate with each other. The way to connect devices, wired or wireless, to the systems will be designed carefully according to the requirements.

8.6.3 Blockchain-based smart home systems

The concept of a smart home [17] is an important aspect of IoT implementation that allows furniture and electric appliances to manage themselves automatically. Among all these projects, none of them have focused on the blockchain application. Therefore it is worth it to find a feasible way to combine blockchain technology with a smart home. Aung et al. [18] proposed a smart home system implemented with the Ethereum blockchain. The system uses the smart contract on the Ethereum blockchain to handle data flow management, data storage, and access control policy. With the smart contract, the utility meter could be tracked and the rules can be set to turn on and off light bulbs and air conditioners to save energy. A private Ethereum blockchain is used to save all the transactions of the smart home system. Owners in each home are provided with different levels of authority to control the accessibility of each owner to the system. However, this is just a proposed idea and it is not implemented in a real testbed environment. Xu et al. [19] presented a PoC of an Ethereum-based system prototype to simulate smart home applications.

8.7 Conclusion and future work

In this chapter we successfully built a PoC system prototype named Uranus using a combination of a private Ethereum blockchain, RPi 3 Model B, the Blynk app, and DHT11 temperature and relative humidity sensors. This system prototype is designed to simulate a smart building application that updates the real-time temperature and humidity to the user and reacts automatically when a certain situation occurs. The sensor data is passed to the smart contract, which is executed on the private Ethereum blockchain. Turning on or off the LEDs as warnings is the reaction for this system when the sensor data exceeds the thresholds set by the user. However, after testing and analyzing the system, some limitations were discovered.

The cost of building this system is quite high; up to $150 for smart home/building applications. This could be solved by finding a more cost-effective substitution for RPi, for example, Arduino, ESP8266, SparkFun, and others. Instead, turning on or off corresponding LEDs is implemented as a simulation process. This could be further improved by researching the process controlling the air conditioner and dehumidifier via infrared technology. An additional process must be conducted, which is keeping enough tokens stored in the Ethereum account of the RPi for the RPi to interact with the smart contract. Since whenever an account wants to send a transaction or call a smart contract on the Ethereum blockchain, the account needs to pay certain token as a gas fee for the transactions to be made successfully.

References

[1] Q. Xu, C. Jin, M.F.B.M. Rasid, B. Veeravalli, K.M.M. Aung, Blockchain-based decentralized content trust for docker images, Multimed. Tools Appl. 77 (14) (2018) 18223–18248.

[2] V.K. Solanki, S. Katiyar, V. BhashkarSemwal, P. Dewan, M. Venkatasen, N. Dey, Advanced automated module for smart and secure city, Prog. Comput. Sci. 78 (2016) 367–374.

[3] H. Sundmaeker, P. Guillemin, P. Friess, S. Woelfflé, Vision and Challenges for Realising the Internet of Things, Cluster of European Research Projects on the Internet of Things, vol. 3, European Commision, 2010, pp. 34–36. no. 3.

[4] Q. Xu, R.V. Arumugam, K.L. Yong, S. Mahadevan, Efficient and scalable metadata management in EB-scale file systems, IEEE Trans. Parallel Distrib. Syst. 25 (11) (2014) 2840–2850.

[5] M.V. Moreno, M.A. Zamora, A.F. Skarmeta, User-centric smart buildings for energy sustainable smart cities, Trans. Emerg. Telecommun. Technol. 25 (1) (2014) 41–55.

[6] B.S.P. Mishra, H. Das, S. Dehuri, A.K. Jagadev, Cloud Computing for Optimization: Foundations, Applications, and Challenges, vol. 39, Springer, Cham, 2018.

[7] S. Chatterjee, N. Dey, A.S. Ashour, C.V.A. Drugarin, Electrical Energy Output prediction using Cuckoo Search supported Artificial Neural Network, in: World Conference on Smart Trends in Systems Security and Sustainability (WS4 2017). London Volume: Springer LNNS Series, 2017. February.

[8] Q. Xu, K.M.M. Aung, Y. Zhu, K.L. Yong, A blockchain-based storage system for data analytics in the internet of things, in: New Advances in the Internet of Things, Springer, Cham, 15 2018, pp. 119–138.

[9] B.B. Mishra, S. Dehuri, B.K. Panigrahi, A.K. Nayak, B.S.P. Mishra, H. Das, Computational intelligence in sensor networks, in: Studies in Computational Intelligence, vol. 776, Springer-Verlag, Berlin, Heidelberg, 2018.

[10] Q. Xu, K.M.M. Aung, Y. Zhu, K.L. Yong, Building a large-scale object-based active storage platform for data analytics in the internet of things, J. Supercomput. (2016) 1–19.
[11] F.M. Chan, Q. Xu, H.J. Seah, Z. Tang, S.L. Keoh, K.M.M. Aung, in: Privacy Preserving Computation in Home Loans Using the FRESCO Framework, Future of Information and Communication Conference (FICC), 2018, pp. 406–414.
[12] J. Paul, Q. Xu, F. Shao, B. Veeravalliy, K.M.M. Aung, Practically realisable anonymisation of bitcoin transactions with improved efficiency of the zerocoin protocol, in: Future of Information and Communication Conference (FICC), 2018, pp. 415–424.
[13] B. Marr, Blockchain and the Internet of Things: 4 Important Benefits of Combining These Two Mega Trends, Available from:https://www.forbes.com/sites/bernardmarr/2018/01/28/blockchain-and-the-internet-of-things-4-important-benefits-of-combining-these-two-mega-trends/#6cf0d2f519e7, 2018.
[14] S. Huh, S. Cho, S. Kim, Managing IoT devices using blockchain platform, in: 19th International Conference on Advanced Communication Technology (ICACT), 2017, pp. 464–467.
[15] Z. Li, Z. Xiao, Q. Xu, E. Sotthiwat, R.S.M. Goh, X. Liang, Blockchain and IoT data analytics for fine-grained transportation insurance, in: IEEE 2nd International Workshop on Blockchain Technologies and Systems (BCTS), 2018, pp. 1022–1027.
[16] C. Lazaroiu, M. Roscia, Smart district through IoT and blockchain, in: IEEE 6th International Conference on Renewable Energy Research and Applications (ICRERA), 2017, pp. 454–461.
[17] N. Dey, S. Fong, W. Song, K. Cho, Forecasting energy consumption from smart home sensor network by deep learning, in: International Conference on Smart Trends for Information Technology and Computer Communications, August, Springer, Singapore, 2017, pp. 255–265.
[18] Y.N. Aung, T. Tantidham, Review of ethereum: smart home case study, in: IEEE 2nd International Conference on Information Technology (INCIT), 2017, pp. 1–4.
[19] Q. Xu, Z. He, Z. Li, M. Xiao, Building an ethereum-based decentralized smart home system, in: IEEE 2nd International Workshop on Blockchain Technologies and Systems (BCTS), 2018, pp. 1004–1009.

Further reading

Q. Xu, K.M.M. Aung, Y. Zhu, K.L. Yong, A large-scale object-based active storage platform for data analytics in the internet of things, in: Advanced Multimedia and Ubiquitous Engineering, Springer, Berlin; Heidelberg, 2016, pp. 405–413.

CHAPTER

Privacy and security of Internet of Things devices

9

Carlos Roberto Moratelli*, Ramão Tiago Tiburski†, Everton de Matos†, Gabriel Portal†, Sergio F. Johann†, Fabiano Hessel†

**Federal University of Santa Catarina (UFSC), Blumenau, Brazil †Pontifical Catholic University of Rio Grande do Sul (PUCRS), Porto Alegre, Brazil*

9.1 Introduction

The modern world is changing quickly for a model where computer networks, especially the Internet, will interconnect billions of electronic devices. These devices will interact with the physical environment bringing their information to the digital media and making physical and digital worlds inseparable. This scenario defines the concept of the Internet of Things (IoT) where the devices are uniquely identified and reachable through the network. An immediate result is the emergence of innovative services and intelligence to the Internet [1]. The possibilities for IoT applications are extensive, for example, agriculture, energy distribution, healthcare, and transportation [2].

In the traditional approach, IoT devices communicate directly with the cloud [3]. However, the amount of data to communicate from sensors to the cloud is prohibitive for some applications. For example, connected cars can generate tens of megabytes each second [4]. Additionally, some applications require fast response, as facial recognition, forcing smartphones to include dedicated hardware for AI processing. Therefore the paradigm of fog and edge layers was added to the IoT architecture [4]. In this concept, sensors will communicate with nearby devices (edge) interconnected by medium range networks (fog). Fog and edge devices can perform data processing and make local decisions avoiding additional communication with the cloud [5]. However, the adoption of this innovative paradigm is followed by significant side effects that must be adequately addressed, for example, cost and security [6].

The innovative benefits of the IoT are followed by an old drawback: digital security risks. In an IoT architecture where edge devices, with significant processing power, will be placed near sensors, the level of resistance against cyberattacks must be considerate. Hence security is the most critical concern for broad IoT adoption since it can bring vulnerabilities to private networks [1]. Consequently, the future internet is a fertile environment for hackers, who use such design flaws to steal

Real-Time Data Analytics for Large Scale Sensor Data. https://doi.org/10.1016/B978-0-12-818014-3.00009-7

sensitive information or to perform denial-of-service (DoS) attacks [2]. Improvements in the security of all devices connected to the Internet are a vital concern for the future of the IoT [6].

While the fog/edge paradigm is considered the future of IoT architecture, the study of security solutions is still in its early stage [7]. Hence this work focuses on presenting four approaches aiming to improve security in edge devices that are complementary and lightweight enough to coexist in the same equipment at an acceptable overhead and footprint cost. These approaches are as follows:

- **Trust execution environment (TEE)**: As many edge devices are placed in public environments with easy access by nonauthorized personnel, it is necessary to guarantee that the running software was not modified or changed maliciously. Secure boot process mechanisms and employing hardware support and cryptography can be used to build a root-of-trust (RoT) environment and, finally, a chain of trust (CoT) with authentication and encryption [6].
- **Security by separation**: The role of separation is to isolate different applications in a spatial and temporal view [8]. Virtualization is the primary technique to provide separation [9], where multiple operating systems or even bare-metal applications can coexist isolated in the same hardware platform. Although virtualization emerged as a typical server-side solution, if carefully adapted, modern embedded systems (ES) can take many of its advantages. Additionally, separation is essential to the maintenance of TEEs.
- **Blockchain for trusted communication**: Communication capabilities are the core of IoT innovation and, at the same time, the main risk for its adoption. A promising technology capable of providing data integrity and trusted communications is the blockchain [10]. It is based on decentralization as a measure of security. It works by keeping a list of records that are linked through digital signature to implement verifiable, secure, and time-stamped transactions between two parties [11]. This technology has several attractive features that make it ideal for future applications, considering all possible security breaches involved in the IoT.
- **Context-aware security**: Traditional software security applications may use an unaware approach to provide security. However, there is a need for a nonstatic method to provide security since the IoT environment is dynamic [1, 12]. Hence context-aware security, which uses context information to provide dynamic security, emerges as a promising approach [13].

Finally, this chapter aims to deliver an overall view of the security concerns and the emerging countermeasures to be adopted in the IoT field. The primary challenge to building a complete security architecture for IoT edge devices is the adaptation of techniques with heavy processing or footprint requirements to constrained ES. For example, while virtualization and blockchain are widely deployed in server farms, its adoption in ES requires in-depth study and innovation. The main contribution of this work is to show how different security techniques can cooperate and be adapted to constrained devices, resulting in a comprehensive security architecture for IoT devices. We show the potential security improvement of each method, explaining how it can be adopted in the embedded field. As a result, we propose a security architecture integrating the four previously mentioned techniques that is affordable for embedded devices.

9.1.1 Chapter organization

This chapter is organized as follows: Section 9.2 highlights the need for security improvements on IoT devices showing recent attacks and redirecting to the possible solution approaches. Section 9.3 demonstrates how digital security can be used to build a TEE. It also provides an overview of RoT, which allows for the secure boot, and consequently the CoT, thus preventing the execution of nonsigned software inside a TEE. Section 9.4 deepens the use of separation as a security countermeasure for IoT devices, evidencing the deployment of virtualization. Section 9.5 demonstrates the applicability of blockchain technology for IoT-trusted communications. Section 9.6 defines context-aware computing and context information, forming the basis for context-aware security and showing how it can be provided in IoT environments. Section 9.7 discusses how these approaches can cooperate to improve security in IoT devices. Section 9.8 presents a literature review and Section 9.9 concludes the chapter.

9.2 The need for security

Security is one of the main challenges for the IoT evolution [2, 14]. The nature of specific applications, for example, the gathering of sensitive information and a large number of connected devices, makes IoT devices a favorite target for attackers. Beyond the actual internet security issues, additional faults are emerging with IoT [6]. Specific applications may represent a severe risk for life or business [15–17]. Imagine a situation where the attacker takes control of an automobile. Other risks include the theft of private information (privacy/confidentiality), modification of the information (authenticity), and malfunctioning of devices (availability and resiliency). Additionally, an affected device may impose security threats for entire private networks, for example, when an affected device, such as an IP camera, is connected to the local network.

Fig. 9.1 presents a taxonomy for the security of IoT devices. We use it to present vulnerabilities, to understand how common attacks can violate security requirements, and to present mechanisms that can strengthen security of IoT devices. We discuss each part of the taxonomy in the next three subsections.

9.2.1 IoT device vulnerabilities

The security of a device can be exploited locally or remotely. Devices can be violated regarding unauthorized access, use, disruption, modification, or even destruction [6, 7]. Most of the attacks aim to disrupt the device's software (e.g., operating system, virtualization, or key storage). Local attacks include unauthorized software updates to add malicious code (e.g., virus or malware). Some devices are installed in open areas without access control allowing physical access to unauthorized people. In this case, it is necessary to guarantee that the device will not be vulnerable to information extraction or software installation. Remote attacks exploit the vulnerabilities of devices connected to a computer network. For instance, attacks exploring backdoor security flaws have been seen in sensitive application environments, like home care [18]. In geographically distributed applications, the devices are visible through the network and use the TCP/IP stack for communication, bringing the well-known internet security issues to billions of devices [2]. Both kinds of attacks require techniques to guarantee authentication, authorization, confidentiality, integrity, availability, nonrepudiation, and privacy preservation [4].

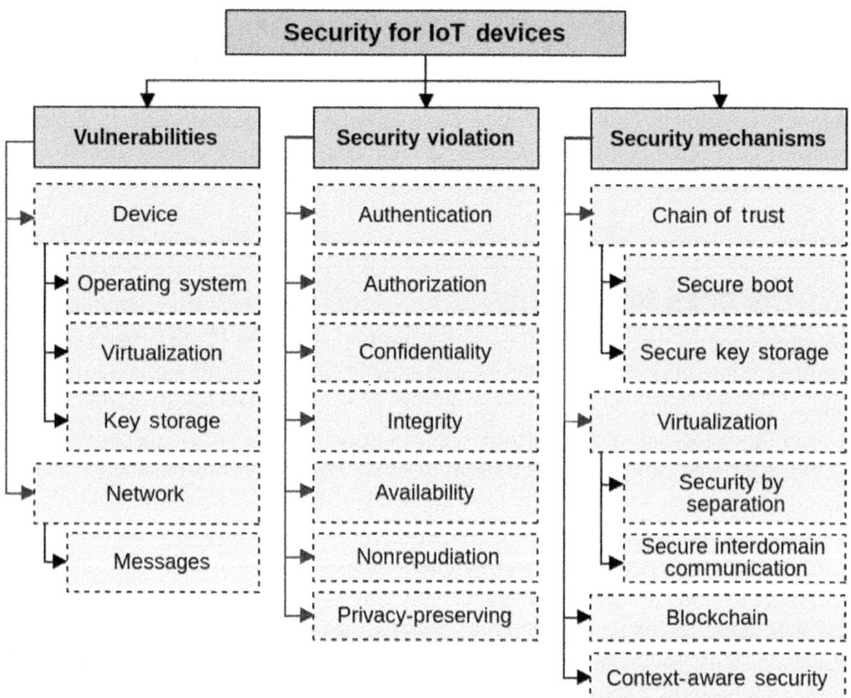

FIG. 9.1

A security taxonomy for IoT devices.

9.2.2 Recent attacks that exploited IoT security fails

Security is emerging as the most significant challenge for IoT devices [2]. The decentralization of data processing and storage to the edge of the network is making them more visible to attacks [4]. According to recent research [1, 6, 7], authentication, confidentiality, and privacy preservation are the primary target of attacks. Default passwords, for example, can be the reason for backdoor violations in IoT devices. Integrity can be violated when the attacker has control of the device, as in intrusion or physical/tampering attacks. Moreover, messages exchanged between devices can also have their integrity broken by attacks on communication channels. Authorization and nonrepudiation can be exploited in privilege escalation attacks, in which the attackers manipulate virtual machines (VMs) doing operations that do not concern them. Finally, availability can be affected by DoS and physical/tampering attacks.

Recently, the Internet was flooded with news of digital attacks that couldn't be imagined some years ago. These events involved, for example, a unique vehicle or thousands of IoT devices working in a coordinated attack. In the past, digital attacks could cost time, money, or loss of information for individuals or companies. But now the severity is much worse. Generally, IoT devices interact with the physical world, building what is known as cyber-physical systems. Attacks to cyber-physical systems have the potential to cause damage in the physical world, putting human life at risk.

On October 28, 2016, a massive attack against the Internet Domain Name System (DNS) infrastructure caused traffic congestion and a complete network standstill in several locations, mainly in the United States. It was used against thousands of IoT devices, such as IP cameras and video recorders, to cause a DoS against Dyn, a company responsible for critical services for websites like Twitter, Amazon, Tumblr, Reddit, Spotify, and Netflix. The attackers employed a malware called Mirai [19] that was released on the Internet just one month earlier in September, 2016, allowing anyone to use it. Mirai searched the Internet for IoT devices with minimal security and infected them and recruited them for DoS attacks. This kind of attack consists of the generation of high loads of traffic against the target. In this case, thousands or millions of low-capacity devices must act together making the target unavailable.

Research has shown many security issues in commercial products, ranging from gadgets to automobiles. For example, researchers from the University of Michigan showed security holes in a family of products called Samsung SmartThings [18]. These products are vulnerable to attacks that could modify their configuration, steal information, or even trigger false fire alarms. The high level of the integration of automobiles with the Internet has made them a target. Researchers managed to break the security of a Jeep Cherokee remotely, taking control of the car's critical control system (e.g., they could turn off the engine while the car was traveling on a highway) [16]. Due to this, Chrysler had to recall 1.4 million vehicles [17].

In a frenzy for the integration of IoT benefits into their products, manufacturers did not pay due attention to security and the potential consequences of a product being hijacked, which can cause considerable financial loss and risks to customers. Now the industry is focusing on improving the security in their products to avoid the same errors of the past.

9.2.3 Opportunities for security improvement

Thus far we have exposed the security problems that affect IoT devices and the potential risks involved. However, there are many opportunities to improve security making the devices resistant against attacks. This section introduces the emerging security techniques that are being developed for IoT devices. We discuss each one briefly in the paragraphs that follow.

The enforcement of the TEE integrity can be done by the integration of secure boot and separation. Secure boot requires a secure bootstrap subsystem in read-only memory (RoM) or a hardware state machine to authenticate the stored program with asymmetric keys [15]. If a modification is detected, the bootstrap process must be interrupted. Security by separation means that a software application cannot influence the execution of other applications on the same hardware, in a similar way that an operating system (OS) performs with its process [8]. However, an OS may not be enough for different reasons. First, some situations require more than one OS running in the same hardware. Second, classical OSs are too big for tiny ES, and the classic real-time operating systems (RTOS) do not perform the required separation and security. Thus virtualization becomes an alternative for bringing separation to IoT devices.

IoT devices are visible in the network exposing them to remotely driven attacks [7]. Blockchain technology, first deployed in cryptocurrencies, can be adapted to the IoT field to provide data integrity and secure interdevice communication [11]. Finally, the dynamicity of IoT environments may require the definition of safe methods capable of automatic adaptation based on its context information, resulting in the enforcement of context-aware security [13].

9.3 Creating and maintaining trusted execution environments (TEE)

A TEE consists in an area of execution of a machine where all code performed and data from the application are verified for confidentiality and integrity [15]. Additionally, TEE requires separation that allows the execution of multiple isolated flows concurrently (see Section 9.4). This section explains the foundations of TEE for boot and execution of trusted code, which consists in a RoT and a secure boot followed by a CoT. Its execution starts during the devices' power-up and persists during all system execution. The following sections describe each one of these elements.

9.3.1 Root of trust

The RoT is an implicitly trusted element that constitutes the foundations for device software integrity [15]. To be trusted, the RoT must be built in a way that it cannot be changed or substituted. However, its implementation may vary from device to device. One approach is to implement the RoT entirely in hardware making it impossible to be replaced. Thus the chip's manufacturer is responsible for its implementation and updates are not possible. Another possibility is implementing it in a bootstrap software on a ROM on-chip. Again, the chip's manufacturer is responsible for the bootstrap software, but it can implement newer versions without significant changes in the integrated circuit architecture. A more flexible approach consists of necessary hardware capable of performing software verification based on a cryptography key stored in a write-once memory. The chip's manufacturer is responsible for providing the support for verification and storage memory. In this scheme, all software stack is the responsibility of the developers and allows for RoT updates even in devices in the field. It is the preferable approach for most of the ES due to its flexibility.

9.3.2 Secure boot process

The secure boot process requires a RoT element to perform code verification in a trusted fashion of the first device's boot stage. Fig. 9.2 describes the general development and device verification processes. First, the developer compiles the source code generating a device's binary image or firmware (circle 1), as in any development environment. Second, the hash digest of the binary image (a numeric code capable of identifying it uniquely) is calculated (circle 2). Third, the hash digest is encrypted by a private key (circle 3) from a public key pair. The result is a signature that must be placed in the device along with the firmware. Additionally, the developer must write the corresponding public key to the device's write-once memory. During startup, the hardware or the bootstrap software recalculates the hash digest from the firmware on storage (circle 4). The last step is to decrypt the signature using the public key (circle 5) and compare it to the firmware digest (circle 6). Both must be the same since different digests indicate that the firmware was changed and the boot process is stopped.

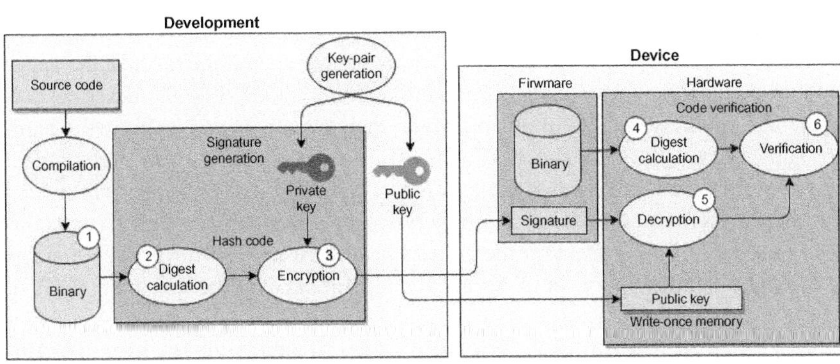

FIG. 9.2

Code signature generation and verification process.

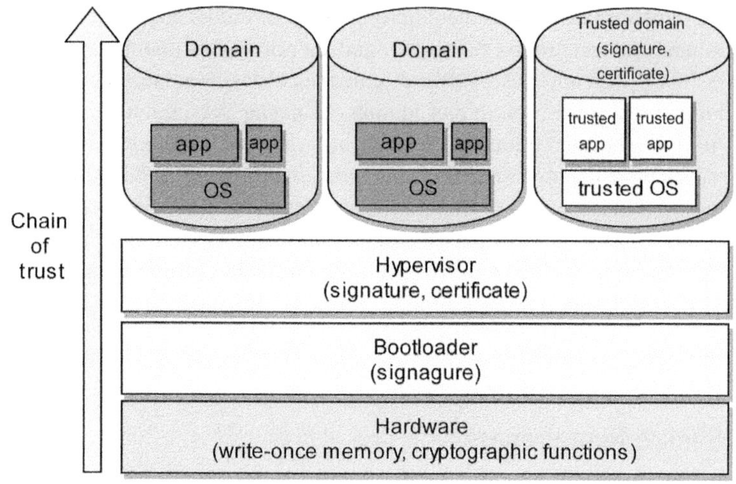

FIG. 9.3

Chain of trust in a virtualized environment.

This secure boot scheme is flexible and preferred among other options. Generally, this scheme is used to verify the bootloader (first boot stage software). Therefore the cryptographic hardware and the bootloader forms the RoT, and it allows developers to implement their bootloader. Section 9.3.3 shows how the bootloader can verify the integrity of the next boot stages. Moreover, as the key in the write-once memory cannot be updated, an attacker is unable to modify the firmware. The weak point of this scheme is the developer's private key since the attacker needs it to generate a compatible firmware.

9.3.3 Chain of trust authentication

As explained in Section 9.3.1, the implementation of a TEE starts with a RoT element [20]. The RoT allows for the secure boot, where only trusted software can be executed on the device's startup (see Section 9.3.2). A trusted software on the device allows for the verification of other software pieces before its execution. This mechanism involves a set of verifications at all layers of the system's software, until the layer of the application, implementing end-to-end security and defining a CoT. Fig. 9.3 describes this scheme in a multilayered environment, which includes a hypervisor. First, the hardware authenticates the bootloader. If successful, it is considered a trusted element and is allowed to verify the next boot stage. In this case, the next boot stage to be trusted is the hypervisor. Once verified, the hypervisor boots up and verifies its domains before their boot. In this scenario the coexistence of trusted and unverified domains is possible, using the hypervisor to enforce the separation, as addressed in Section 9.4.

9.3.4 Physical unclonable functions—An alternative for key storage

Physical Unclonable Functions (PUFs) are hardware mechanisms that allow for low-cost authentication and key-generation applications [21]. A PUF is implemented using the natural physical variations emerged from the semiconductor manufacturing processes, which let signatures that are impossible to clone. PUF uses simple digital circuits since they are easy to fabricate and consume less power than nonvolatile memories [21]. PUFs can be used to replace the need for key storage since the key generation occurs at runtime. The software reads the PUF and uses its unique fingerprint as a seed for the cryptographic algorithms. Therefore attacks to extract the key or other secrets must occur during normal execution, making them more difficult. There are several kinds of PUFs, a typical example is the SRAM-based PUF or SRAM-PUF, which uses the power-on state of the SRAM, unique for each device, to generate a genuinely unique identification number and cryptographic keys.

9.4 Security by separation

Security by separation is a classic approach for protection that proved to be reliable over time [8]. It can be implemented using different physical devices or logically, with the goal of restricting the availability and access to resources, applying policies of use. Separation by different physical devices requires resource duplication implying higher costs. For example, keeping two software functionalities in different memory and processors. Logic separation can be achieved by virtualization, which is capable of simulating a platform, allowing multiple domains or applications to share the underlying hardware and avoiding resources duplication. Additionally, the characteristics of a given application may require some level of interaction, such as data dependencies or synchronization, meaning the involved entities should not be wholly separated. Most IoT solutions will require cooperation with sensors and transmission over the network making complete isolated systems less attractive. Thus this section focuses on systems that use separation for security improvement while still allowing interaction and communication according to a defined policy.

9.4.1 Lateral movement

Software complexity is continually increasing and has reached a point where it is difficult to fix all vulnerabilities [22]. Additionally, developers of IoT devices are stimulated to minimize hardware costs integrating as many functions as possible in the same platform [23]. Another point to be considered is the time to market, which forces manufacturers to speed up their development by overlooking the test step. In this scenario, it is most likely that security flaws will appear during the product's lifetime.

In a compromised system, the attacker will try to spread its attack to other subsystems, taking control of all possible functionalities. This is called lateral movement and is a widely used tactic. An example of this tactic is in the attack of a corporate network. An initial entry point could be a malware sent by email attachment or some other way. Once inside of a machine, the attacker will try to move to other machines with the aim of reaching the database or other desired target. In the case of the Jeep Cherokee discussed earlier, the researchers posing as attackers [16] (see Section 9.2.2) took advantage of lateral movement. They began by accessing the vehicle's entertainment system and moved on to critical functions such as braking and transmission. We know that it is almost impossible to fix all security breaches in complex systems, but we can isolate the flaws restricting the lateral movement inside a system. Unfortunately, most of the manufactures fail to isolate the internal system functionalities, exposing all software and hardware in the case of an attack.

Due to the costs, engineers will keep aggregating functionalities to the same hardware platform. As physical separation will not be implemented in most IoT devices, the logical separation can help to achieve cost restrictions and the desired level of security.

9.4.2 Implementing separation

As mentioned, the quick advancements of semiconductors allowed more functionalities to coexist on the same platform, allowing applications to share processor, memory, and other platform resources. Additionally, the adoption of multicore processors strengthens such a trend. Virtualization is the technology capable of performing separation, enforcing the software to execute in different logical domains utterly unaware of other instances, even when sharing hardware resources [20]. Both techniques can be applied to multicore or even single-core platforms. This section explains how spatial and temporal separation can be achieved.

9.4.2.1 Spatial separation

Spatial separation is the ability to isolate applications (data and code) to share the same memory [8]. Memory sharing is a practical solution to reduce costs and to simplify software implementation. The memory is connected to a processor through data, address, and control buses, and applications share the same physical address space. Thus communication can be implemented using shared memory, making it faster than when using message-passing mechanisms. However, without proper separation, a running application can have access to all processors' address space, that is, a malicious or malfunction code can touch any memory location. To provide memory separation, the processor needs to be equipped with a hardware block, which provides virtual memory abstractions to the system, called a Memory Management Unit (MMU). Initially, the MMU was present in desktop and server processors, being mandatory to the execution of complex and stable operating systems. However, nowadays a large number of embedded processors have some level of memory management. In these processors, the generated memory addresses are called virtual addresses (VAs) and need to be translated to physical addresses (PAs) by the MMU, which is managed by the host OS or the virtualization layer (when present). The OS keeps a table, called a page table, which maps VAs to PAs for each performed process. The MMU performs translations with a Translation Lookaside Buffer (TLB) search for each virtual address, converting VAs to PAs. TLB entries keep the most frequent translations, avoiding page table walks for each memory access. When the translation is not succeeded (occurrence of a page fault) a trap is issued to the OS. Thus malicious or malfunctioning memory access is detected, and the responsible process can be ended. This scheme is shown in Fig. 9.4A.

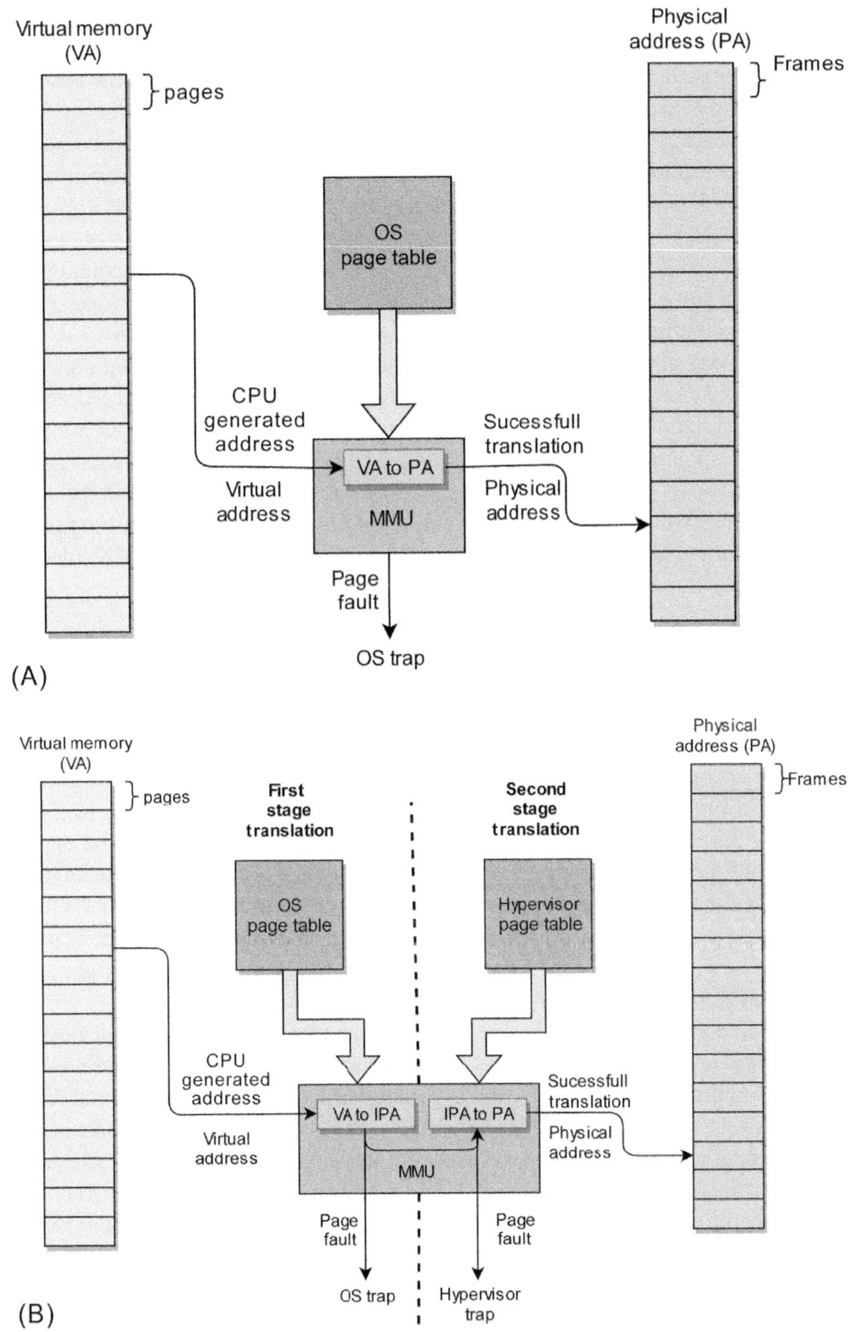

FIG. 9.4

Memory address translation using one- and two-level MMUs: (A) Address translation using one-level MMU and (B) Address translation using two-level MMU.

When the virtualization layer is present, it must keep the separation between VMs or domains, while the guest OS (virtualized OS running in a VM) must be able to maintain separation between processes. This is possible by creating a new abstraction level in the memory address translation. VAs are translated to Intermediate Physical Addresses (IPAs) that are managed by the hypervisor (the implementation of a virtualization layer). However, the efficient management of IPAs may require additional MMU hardware features. Without proper hardware support, the IPA management needs to be emulated by the hypervisor, using a technique known as shadow page table, which keeps the relation between IPA and PA in an intermediate page table. Virtualization on processors without proper support brings additional complexity, and sometimes prohibitive overhead, to the resulting hypervisor.

The adequate support for memory translation adds a two-level translation to the MMU, as seen in Fig. 9.4B. The MMU is responsible for translating from IPA to PA without software intervention. The hypervisor still manages its page table by mapping the correspondence between IPA to PA, but the hardware performs the actual translation. Thus the guest OS leads the first stage of address translation in the same way as on a nonvirtualized system, with the support of the TLB. The resulting physical address is a combination of first-stage VA and second-stage IPA. This mechanism decreases hypervisor intervention and software complexity.

9.4.2.2 Temporal separation

Temporal separation is the ability to guarantee the fair distribution of the processor's time between applications according to their execution priorities [8]. Differently from spatial separation, temporal separation is performed by software and does not require special hardware features, despite some processors implementing functionalities to improve performance. This approach, also known as timing sharing, emerged with the first OSs as a solution for improved processor utilization and to allow multiple applications and users to share the system's resources simultaneously. In ES, temporal separation is a concern for two different reasons. First, many ES have severe time constraints and the temporal separation must address this adequately. Second, a malicious or malfunctioning use could deteriorate or even avoid the execution of other software by hogging the CPU.

The implementation of temporal isolation is performed by a software block called scheduler that is implemented in both OSs kernel and the hypervisor, the case of virtualized systems. Hypervisors create a CPU abstraction, known as virtual CPU (VCPU) that is bound to the VMs. The VCPU is a software representation of the physical processor and allows the execution of time multiplexing. Thus the scheduler performs an algorithm that must choose between the available VCPUs the next one to enter in execution. There is a wide variety of scheduling algorithms and many works have addressed the hypervisor scheduler as a way to improve temporal separation, as seen in Ref. [24].

Finally, system interrupts are a concern for temporal separation not only because of their interference on the correct timing execution, but also because they can be used to disrupt the normal operation. For example, an attacker could inject a massive number of interrupts causing DoS, since the system will be busy handling them. Thus the interrupt subsystem must be carefully implemented as part the of the hypervisor scheduler. In this way, the hypervisor can implement interrupt delivery policies according to the application needs and time constraints. In the case of interrupts affecting the timing, the hypervisor can mask them temporally to prioritize critical functions.

9.4.3 Virtualization

Virtualization means the creation of an environment or VM that acts like the real target hardware from the software or user point of view [8, 9, 20]. The VM is implemented as a combination of real hardware and software aiming to execute applications [25]. The virtualization layer is often called a virtual machine monitor (VMM) or simply hypervisor. Fig. 9.5 depicts the virtualization taxonomy. VMs designed to support different Instruction Set Architectures (ISAs) need to emulate all software (OSs and applications). Thus they are also called whole-system VMs [25]. There are important applications for this kind of VM. For example, an ARM platform including the processor, memory, and peripherals can be fully emulated on an Intel x86 processor. Thus the designers can test their software without the target hardware. Additionally, legacy software can be supported by newer platforms without requiring engineering work to port. An example of a system VM for different ISAs is the QEMU [26]. QEMU is an open-source application that can be used either as an emulator or virtualizer. As an emulator, it can run OSs and programs compiled for different ISAs. When used as a virtualizer, it can execute the guest OS code directly on the host CPU.

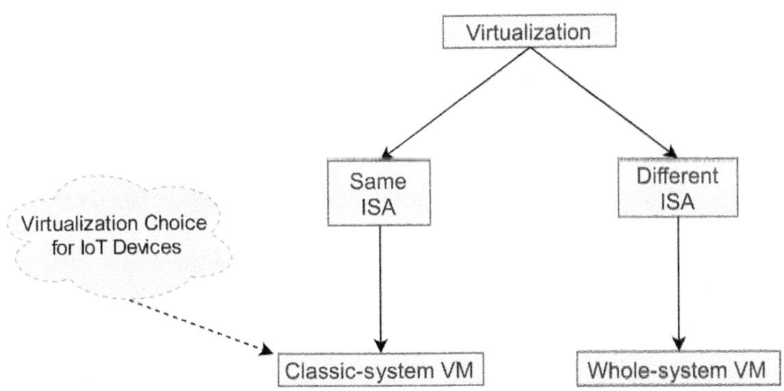

FIG. 9.5

The division of virtualization in classic-system VM and whole-system VM.

VMs designed to the same ISA, also called classic-system VMs, were developed during the 1960s and they were the first application of virtual machines. At that time, mainframe computer systems were large and expensive. Thus computers needed to be shared among users. Different users sometimes required different Oss, creating the first practical application for virtualization. With the popularization of PCs, interest in classic-system VMs was lost. However, during the 1990s with the appearance of server farms, the classic-system virtualization became popular once again. In this case, servers are shared among several users, and VMs are a secure way of partitioning the user's software. Classic-system VMs have a lower performance impact when compared to whole-system VMs because the hardware directly executes the VMs' instructions. Recently, the interest in classic-system VMs for ES has increased. However, classic-system VMs as applied to server virtualization do not fit directly with ES. Their intrinsic characteristics (memory constraints, real-time execution, and security needs) motivated the study of new approaches for virtualization on IoT devices. Additionally, there are two different techniques to build a virtualization layer: implementing hypervisors type-1 or type-2. Type-1 hypervisors (Fig. 9.6A) perform directly on top of the hardware being used in server virtualization, and they are the preferred choice for ES since they do not require the additional overhead of a host OS. An example of this kind of hypervisor is the Xen [27]. Type-2 hypervisors (Fig. 9.6B) are performed on top of an OS, and they are the preferred choice for home and office virtualization since they can perform guest OSs on the side of a user's native applications. An example of a type-2 hypervisor is the Kernel Virtual Machine (KVM) [28].

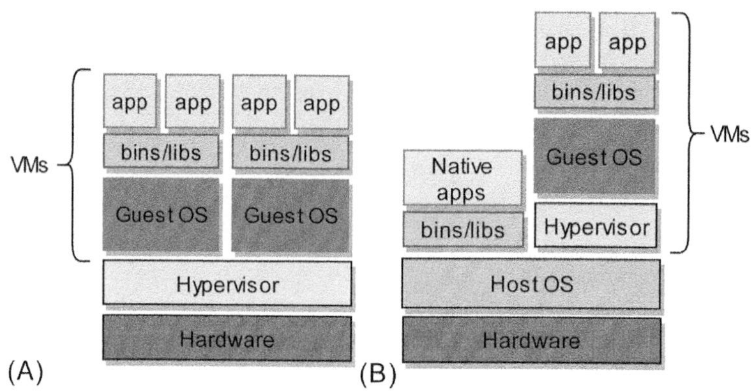

FIG. 9.6

The two different kinds of hypervisors: (A) Hypervisor type-1 and (B) Hypervisor type-2.

9.4.3.1 Embedded hypervisors

Although server virtualization is a well-known and mature technology widely applied for commercial use, ES virtualization is still being studied and developed. The main restriction to broader adoption of virtualization for ES is that their requirements differ from server and enterprise systems [16]. Most notably, time constraints and limited processing power and memory size are the main concerns around ES virtualization. Thus virtualization, as deployed in the enterprise environment, cannot be directly used in ES [8].

Many studies are proposing different techniques for ES virtualization. A common approach is to adapt hypervisors widely used in server virtualization to ES. The main concern about general-purpose hypervisors is the absence of suitable support for real-time and memory requirements, making them incompatible or useless for some embedded devices. Despite these limitations, researchers are continually working to improve the embedded support of open-source hypervisors for server virtualization, like XEN and KVM [24, 29–31]. However, the distinct characteristics of ES have motivated the appearance of hypervisors specially designed for embedded virtualization. Among the goals for the development of embedded hypervisors, two of them are frequently addressed: (1) to keep low memory requirements and (2) some level of support for real-time applications. Additionally, the diversity of ES and their applications motivated the development of many different embedded hypervisors, like SPUMONE [32], XtratuM [33], M-Hypervisor [34], Xvisor [35], the family of L4 hypervisors [36, 37] and the Hellfire Hypervisor [8]. All of these are type-1 hypervisors, as addressed in Section 9.4.3.

9.4.4 Enabling techniques and technologies for virtualization

There are three main techniques adopted for hypervisor implementation: (1) para-virtualization, (2) full virtualization (FV), and (3) hardware-assisted virtualization. These techniques can be utilized separately or through a hybrid approach, combining two or more of them.

FV presupposes a processor that supports the implementation of a hypervisor capable of delivering virtualization without any modification on the guest OS. However, such a technique requires that the processor architecture fits the Popek-Goldberg theorem [38], which demonstrates the ISA requirements. This allows the virtualization of unmodified guests dispensing additional engineering work. However, the hypervisor must emulate privileged instructions that can cause performance penalties, as the time needed to emulate instructions using the trap mechanism is thousands of times slower than the execution of a native instruction.

Hypervisors for processor architectures that do not fit the Popek-Goldberg theorem require different approaches. A solution is to modify and recompile the guest OS before its installation in a VM, resulting in para-virtualization. These modifications substitute the sensitive instructions (instructions that modify the processor's behavior) or other OS parts, like the I/O subsystem, by hypercalls. Hypercalls are services provided by the hypervisor and invoked from the guest OS. They are equivalent to system calls (syscalls) in an OS. Additionally, PV can be used to extend the hypervisor capabilities, such as the implementation of inter-VM communication. Another application for PV is to improve performance in systems where fully virtualized hypervisors suffer from performance issues.

The need for better-performing and simpler hypervisors stimulated processor manufacturers to adopt hardware support, resulting in hardware-assisted virtualization. The hardware support can be implemented in a variety of mechanisms. For example, MMUs with support for two-level memory address translation (as seen in Section 9.4.2.1); different levels of privileges allowing hypervisor, OS kernel, and application to perform with different permissions; and interrupt virtualization for better hypervisor control over the system I/O. Therefore the processor's manufacturer determines the level of hardware-assisted support based on the market niche, required features, and costs.

9.5 Blockchain for trusted communication

With the increase in the number of connected devices to the Internet it is essential to find a way of ensuring data integrity and trusted communication for IoT environments [22, 23]. Connected devices may spread sensitive personal data and reveal the behaviors and preferences of their owners. Thus critical privacy issues may arise [7]. Blockchain has emerged as an essential technology to ensure trusted communication between two parties [39]. With blockchain, applications can operate in a decentralized way, without the need of a central authority [11]. This section provides a theoretical and technical background of blockchain technology.

9.5.1 Definition

A blockchain is a distributed data structure that is replicated and shared among the nodes of a network [10]. It has been used in a lot of IoT applications because it successfully achieves secure transactions between nodes [11]. Some concepts around blockchain are as follows:

- **Transaction**: This represents a transference of value between two nodes. Each transaction is processed and recorded in the current block.
- **Blocks**: A block is responsible for recording transactions. A timestamped candidate block is verified by the nodes to check if it should be added in the chain or if it should be discarded. Every time a block is completed, a new one is generated. A blockchain can have many blocks, which are linked to each other as presented in Fig. 9.7.
- **Nodes**: The entities in a blockchain participate in a P2P network, which consists of a series of devices that act as nodes. A new message in the network is propagated between all nodes. There is no way to track who added the message on the network; we can only verify its validity.
- **Consensus mechanism**: Nodes of a blockchain network use a consensus mechanism to agree on the transactions and in the order that they are listed on the block [10].

Blockchain nodes use a pair of public/private keys to interact with each other in the blockchain network. Private keys are used to sign their transactions while the public keys make them addressable on the network [10]. Asymmetric cryptography through the Elliptic Curve Digital Signature Algorithm (ECDSA) [40] provides nonrepudiation, authentication, and integrity to the network. On the other hand, hash algorithms

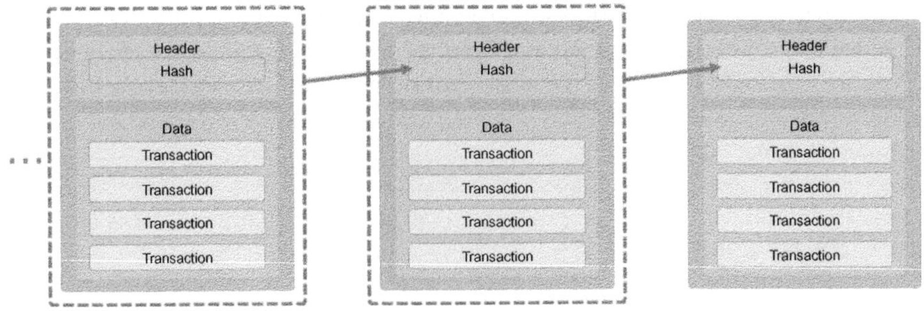

FIG. 9.7

Blockchain architecture.

such as SHA256 [41] can be used to generate the hash of each block, which is generated over the block's contents and its header. Therefore, as seen in Fig. 9.7, the described behavior creates a chain of hashes that make it impossible to change the data already added. Each block has a list of transactions and a hash to the previous block (except to the first block, not pictured). In case of data modification, all hashes from the next blocks would become invalid.

9.5.2 Consensus mechanisms

The blockchain network is usually a set of nontrusting nodes sharing a database with a nontrusted middleman [39]. Thus the question is: what constitutes a valid transaction? In order to help the network reach consensus, each blockchain needs to establish specific rules that each database transaction should conform to [10, 11]. The rules are programmed into each node and are used to decide if an incoming transaction is valid or not [10]. Trust is achieved as an emergent property from the interactions of different participants in the system [11]. Some examples of consensus algorithms are presented next [10, 11, 42].

- **Proof of work (PoW)**: Consists of solving a mathematical puzzle, also called mining, attempting to cryptographically tag a block as valid. The miners in the network compete between each other to solve the puzzle and the one that solves it first is awarded with an amount of value, in cryptocurrency applications, usually coins.
- **Proof of stake (PoS)**: Consists of establishing and assuming that the most influential nodes are honest and will not try to tamper the network. Thus these nodes have the authority to determine and sign the valid blocks.
- **Delegated proof of stake (DPoS)**: As in PoS, DPoS assumes that most influential nodes are honest and will not try to tamper the network. However, although they have authority to validate and sign blocks, the influential nodes are elected as witnesses to testify that some other random node will correctly validate and sign the blocks.

- **Practical byzantine fault tolerance (PBFT)**: Consists of a voting system where some nodes are elected as voters, which can be composed by a slice or all of the network nodes. Thus an election is required to add a new block to the chain.

9.5.3 Smart contracts

A smart contract is defined as a computerized transaction protocol that executes the terms of a contract [10]. The idea is translating contractual clauses into the code and embedding them into nodes that can self-enforce to minimize the need for trusted intermediaries between transacting parties [11, 43].

A smart contract is triggered by addressing a transaction to it. Then, it executes automatically and independently in a prescribed way on each blockchain node [10]. Smart contracts allow for general purpose computations occurring on the chain. The contract has a state and can take control over assets on the blockchain [42]. It allows for expressing business logic in code. It operates as an autonomous actor whose behavior is entirely predictable (it is deterministic).

9.5.4 Blockchain for IoT devices

An IoT system is usually composed of a set of devices that can be based on a blockchain network [42, 43]. The blockchain has the main role of registering and authenticating all operations performed on IoT devices' data. Each operation on data (deletion, modification, creation) is registered in the blockchain; this ensures that any abuse on data can be detected. Also, access policies can be specified and enforced by the blockchain, preventing unauthorized operations. The blockchain can improve security in IoT environments through the following features [44]:

- **Immutability**: The cryptographic chain implemented with the hash algorithms grant that the data inserted in the blockchain became unchangeable.
- **Integrity**: The immutability achieved by the blockchain grant that every data stored be intact.
- **Authorization**: The digital signatures provided by algorithms such as ECDSA are used to ensure that only authorized entities interact with the due content.
- **Fault Tolerance**: To interact with the network each entity needs to agree with the utilized mechanisms.
- **Transparency**: The privacy model adopted enables the data in the blockchain to be completely transparent without compromising the user's identity.
- **Availability**: The decentralized nature of the blockchain allows that if one or more nodes leave the network, the remaining nodes still keeping the network running.

- **Consistency**: The consensus algorithm adopted by the nodes ensures that only valid transactions be included in the blockchain network.
- **Privacy**: The cryptographic model implemented by digital signature algorithms ensures that the user identities are kept in a private domain even in public networks.

Although such security features can improve security in IoT environments, there are challenges to be overcome [43]. Processing power and storage can be challenging in resource-constrained devices. Hence recent works have explored new ways of improving the resource consumption of IoT devices when implementing a blockchain [45, 46]. Eliminating PoW and using an alternative like distributed trust is being discussed as a way to lower the cost of the system's resources without compromising the security and reliability of the transactions [46]. Also, to develop specific architectures to distribute the computing proportionally between the entities with high and low resources is frequently discussed to unflood the restricted devices of expensive tasks [45].

9.6 Context-aware security

The use of context information to provide security decisions has become a trend in IoT environments [47]. This technique is called context-aware security and stands out from the traditional security methods by the use of dynamic IoT device data and events instead of static parameters [13]. Context-aware security can also be used to double-check and prevent possible security issues. This section defines context-aware computing and context information, which is the basis for context-aware security, and presents ideas for how context-aware security can be provided in IoT environments.

9.6.1 Context-awareness and IoT

The IoT is a dynamic environment and therefore it is essential to sense the environment and turn it into valuable information in order for IoT entities to make accurate decisions [48]. Context is considered as any high-level information that can be used to characterize the situation of an entity (e.g., IoT device, person, or place) that is relevant to the interaction between a user and an application, including themselves [49]. According to Perera et al. [48], the context information can be divided into two types:

- **Primary context**: Any high-level information that can be retrieved by a reasoning processing (e.g., business rules, ontology, probabilistic) from raw data, and without performing any sensor data-fusion operations. An example of a primary context information is the inference of a location (e.g., city, state) based on GPS sensor readings (see Fig. 9.8). Primary context may vary depending on the producer.

- **Secondary context**: Secondary context is characterized by the outcome obtained by the computation of primary context. It can be computed by using sensor data-fusion operations (e.g., aggregation) or data-retrieval operations (e.g., web service calls). An example is the identification of the distance between two sensors by applying sensor data-fusion operations on two raw GPS sensor values.

A system is considered context-aware if it uses context to provide relevant information and/or services to the user, where relevance depends on the user's necessities [48]. In this way, an IoT environment requires a context-aware mechanism to be aware of the environmental situation in order to help the user in the most useful way. In various cases, context-aware becomes a feature of IoT systems [12].


```
{
    "provider_id": 9588,
    "latitude": [-29, 58, 26, 4],
    "longitude": [-51, 11, 41, 6],
    "location_city" : "Porto Alegre",
    "location_state" : "RS",
    "location_country" : "Brazil",
    "time" : [08, 07],
    "period" : "PM",
    "date" : [10, 10, 2018],
    "info" : "User 704"
}
```


FIG. 9.8

Context information of an entity's location in JSON format.

9.6.2 Context-aware security mechanisms

Most of the context-aware systems produce the context information and use it only to notify the user/application about an event. Some systems may use it for a different kind of processing, like decision-making. However, there are only a few systems that provide context-based security decisions by using context information, which is considered a gap in the area [12, 13, 48, 50]. Context-aware security is defined as the high-level information (i.e., context) acquired from the environment (e.g., users, applications, places) that can be used in the security mechanisms [47].

It can also be defined as a situation where a security solution considers a set of information (context) while making a specific security decision. For example, security mechanisms may adapt authentication methods depending on a user's location [51]. The context-unaware systems may be outdated for the IoT once these environments have dynamic characteristics. The security mechanisms will need dynamic features as well. Thus the context information can be used for refined security and privacy provisions by adjusting security parameters in execution time. For example, the system access of a user may vary depending on its location. The context-aware security mechanisms can even sometimes omit authentication because their context is trustworthy by itself (e.g., access from inner company network).

Using context allows system administrators to make more fine-grained security rules, which would be otherwise tangled with multiple rules making them unsustainable for maintenance [52]. Contextual information can be integrated into various security mechanisms such as authentication, access control, and encryption, among others [51]. Next we present how context-aware security can be used in IoT environments [51, 52].

- **Authentication**: Traditional authentication mechanisms are context-insensitive (i.e., they do not adapt their security policies to a changing context). Nowadays, passwords are the most common form of authentication. However, passwords are also a significant source of vulnerabilities, as they are often easy to guess, reused, forgotten, shared with others, and susceptible to social engineering [53]. Thus the use of context information to make authentication processes stronger is a suitable choice for such dynamic environments.
- **Authorization and access control**: Even though authorization and access control represent two different areas, most approaches present solutions for both in one package. Many existing systems work with the concept of allowing and denying access by policies. The access can be granted for different situations, such as enter in a network, access data, modify devices, and others. "To allow" means that an entity (e.g., user, device, application) has the credentials that match with the stored ones. "To deny" means that the credentials do not match. This kind of traditional access control is considered static because the credentials to grant access will never change. However, in IoT environments, it is important to consider dynamic aspects when granting access to an entity. The context information can be used to provide flexible security policies. For example, the access can be granted in a predefined time of the day if the user is in a defined location, and access can be denied if these contextual conditions are not met. There are many principles for authorization and access control, including role-based access control (RBAC), rule-based access control (RAC), attribute-based access control (ABAC), and many others. Also, some authors merged principles to create a new access control called, context-aware role-based access control (CARBAC), to meets with their necessities [54].
- **Privacy-preserving**: A user's daily activities can generate sensitive information (e.g., places visited, health conditions). Thus it is mandatory for IoT applications to keep such kind of information private [55]. Location-based service (LBS) provides a user with customized services by using the context information, such as the user location and nearest restaurants/hotels, which are retrieved from a database stored remotely in the LBS server. LBS not only serves individual users, but also plays an essential role in public services like transportation, public safety, and disaster management. With an increasing number of mobile devices with GPS technology built in, LBS solutions have experienced a significant growth over the past several years. Despite the benefits provided by LBS, users may not be willing to offer their current location to the LBS server due to concerns of location privacy

[56]. The objective of a privacy-preserving LBS is to protect the privacy of a user's location while maintaining a high level of LBS accuracy. The context information can be used to determine when or when not to keep user information private.

FIG. 9.9

Providing context-aware security for IoT environments [13].

Context should be a first-class security component to drive the behavior of IoT devices and thus adapt the security decisions to the context in which transactions are performed. At the same time, context information should be managed considering privacy and security interests. In particular, current IoT devices can obtain context information from other entities of their surrounding environment as well as provide contextual data to other smart objects [50]. These communications can be performed through lossy networks and resource-constrained devices, which must be secured by suitable security mechanisms and protocols.

Fig. 9.9 presents a potential attack and how context-aware security deals with this issue. Fig. 9.9-A shows an IoT application scenario that performs some decision-making. For instance, **when** the temperature is greater than 27°C **and** the user location is defined as *home*, **then** a specific window of the house is open.

In Fig. 9.9-B, the temperature is still the same, but an attacker has changed the user location. The attacker may change the user location to *home* in order to open the house window (i.e., get access to a specific resource).

The context-aware security tool takes care of such issues as demonstrated in Fig. 9.9-C. Every time that a crucial security decision has to be made by the application, it checks the information with the context information repository. The repository contains the last context information sensed (i.e., near real time). Also, it has a history with the context sensed in a determined situation, for example, every day at this time the user tends to be at his office, so it is unusual to his location be *home*, therefore access will not be automatically granted.

The scenario presented in Fig. 9.9 shows how context-aware security works for authorization and access control. However, we can also consider that the user must register to the system (i.e., authentication), and this registration may consider the user's context, so he will only be apt to authenticate when being in a determined location (i.e., at home). Regarding the privacy-preserving context, the user may choose to not share data with third-party apps when being in a specific context (e.g., at home, at a specific time of a scheduled meeting).

The core process for provisioning context-aware security is called reasoning. It infers a security decision based on the environmental data/context. Most of the systems that implement the context-aware security feature use rules for the reasoning process. Rules are considered a low-consumption technique since they use few resources for inference processing [48]. The rules can be used for authentication, authorization and access Control, and privacy preservation. Besides this, the entity that implements context-aware security must hold or have access to a database containing past data/context information for checking and inferencing. The context information repository represents the processing of searching for historical data/context, as shown in Fig. 9.9.

9.7 Technologies integration in a comprehensive security architecture for IoT devices

The preceding sections showed that the IoT edge devices introduce potential failures in private networks. Additionally, their intrinsic characteristics require different approaches to cover all vulnerabilities and, no less important, these solutions must cooperate to improve overall security. This section discusses how an IoT device can take advantage of the four different security mechanisms to refine its protection: (1) TEE (see Section 9.3), (2) security by separation (provided by virtualization; see Section 9.4), (3) trusted communications (provided by blockchain; see Section 9.5), and (4) context-aware security (see Section 9.6). First, we identify the target devices of our solution. Second, we present our proposed solution, and finally, we explain how our solution could be applied in real life.

9.7.1 The target devices for protection

Fig. 9.10 describes fog and edge computing architecture. According to the OpenFog Reference Architecture [5], the cloud works as a coordination layer responsible for making general decisions (e.g., security and storage), putting almost all the

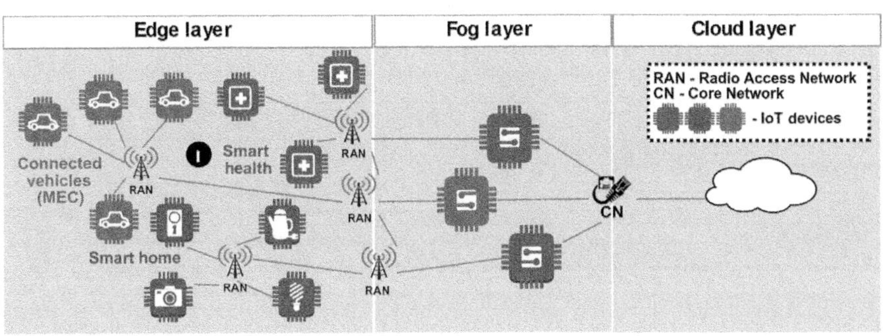

FIG. 9.10

Fog and edge computing architecture.

processing, control, and security to edge and fog layers. The core network (CN) connects the fog to the cloud. The fog layer is composed of powerful devices that should make decisions, store data, and manage edge devices connected through the radio access network (RAN), for example, WLAN, WiMAX, and cellular networks. The edge layer is composed of IoT end-devices. They are usually resource-constrained devices that can preprocess data and make decisions at the edge layer [4]. This makes them the most critical part of the architecture and the starting point for security [6, 7]. Our proposed solution consists of integrating different mechanisms to leverage the security in these devices.

9.7.2 The security architecture

Fig. 9.11 presents the integration of the four mentioned security technologies, showing how they can coexist in the same device and considering fog and edge computing. The first step to protect the device is to establish a CoT (Fig. 9.11-A), starting with the hardware and moving to the highest level of software. An IoT device must be designed to boot up only if the first piece of software to execute is cryptographically signed by a trusted entity (e.g., device vendor). Specialized hardware, as write-once memories, can be used to store cryptographic keys and perform the signature verification of the first stage bootloader (Fig. 9.11-B). This scheme enables a RoT that, anchored in the hardware, makes tampering more difficult and is used to ensure the integrity and authenticity of all running software. Once RoT has been established, the following software component should make similar integrity checks with the hypervisor in the boot chain (Fig. 9.11-C). If successful, then the same process will take place in the next boot stage of the CoT until the software stack is fully protected. The primary purpose of the CoT is to deliver a verified software stack to the runtime environment, helping to prevent physical/tampering attacks.

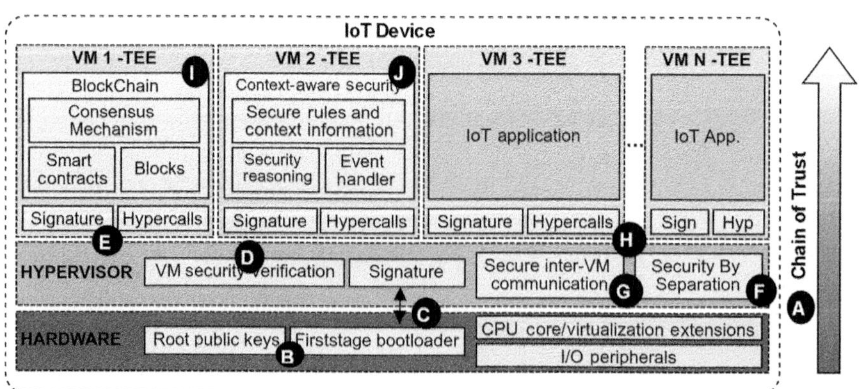

FIG. 9.11

A comprehensive security architecture for IoT devices.

With the hypervisor authorized, it must verify the integrity and authenticity of the VMs. This is possible because the hypervisor is securely anchored to the hardware, being part of the CoT, and trusted for all operations (Fig. 9.11-D). Additionally, a flexible security architecture must provide mechanisms for each software provider to be responsible for the integrity of its software. The integration of secure boot and virtualization mechanisms allows for the software providers to sign their VMs independently, that is, each VM has a signature header (Fig. 9.11-E). Any provider can add its public key to the signature header allowing the hypervisor to check the VMs individually, therefore ensuring code integrity, authenticity, and nonrepudiation. Hence our proposed architecture guarantees that the firmware was not modified (integrity), the VM is from the provider it claims to be from (authenticity), and if a VM causes a malfunction, the responsible party cannot deny it (nonrepudiation).

Once the CoT is complete, each VM is a TEE, and the virtualization layer is responsible for preventing possible runtime violations by keeping the integrity of the execution environment. As presented in Section 9.4, security by separation uses spatial and temporal separation to avoid software defects in a part of the device to propagate to an adjacent part of the physical platform (Fig. 9.11-F). The hypervisor's spatial separation, enforced by the hardware, improves the confidentiality. If an application attempts to access a memory region of any other application or even peripherals, it will be stopped by the hypervisor. Temporal separation guarantees the fair use of the processor's time. Security by separation improves security over major attacks, such as DoS, backdoor, hidden-channel, intrusion, and privilege escalation [7]. Additionally, VMs require some level of interaction between them and the hypervisor itself. Thus a secure inter-VM communication mechanism should be available in the hypervisor (Fig. 9.11-G). It must be implemented as para-virtualized services, that is, using a well-defined hypercall API (VM's calls to the hypervisor) as presented in Fig. 9.11-H. Thus the hypervisor can work as a communication arbiter, copying messages from the sender to the destination VM.

Virtualization also brings an advantage for keeping the integrity of TEEs. It allows the hypervisor to monitor the behavior of the VMs, detecting malfunction caused by software errors or attacks. There are three ways of detecting a compromised VM: (1) a VM tries to access memory outside the address space defined at design time, (2) a VM invokes hypercalls that should not be called, and (3) a VM does not periodically reset a watchdog. If the system detects one of these situations, the VM will be restarted. Hence two things can happen: (1) if the VM code were compromised after the deployment, its digest signature would not match during the CoT phase and it will not boot up and (2) if an attacker was exploiting a vulnerability based on the malfunction, it will boot up and run as expected. However, the hypervisor should emit alerts about the reset activities enabling the developers to investigate the causes.

At this point, an IoT device is still vulnerable to attacks that explore the integrity of messages exchanged with other devices of the network. As shown in Fig. 9.10, an IoT device is usually part of a set of devices that interact with each other to provide relevant solutions for people, such as smart homes, smart transportation, and smart health [23]. The deployment of a blockchain would improve the security of such IoT environments regarding message integrity and device authorization, preventing severe attacks such as message modification. Each IoT device should host the consensus algorithm and the blockchain (including smart contracts). In our proposed architecture, this can be done in a VM (Fig. 9.11-I). As a protected VM, it will inherit the benefits of security provided by virtualization and TEE mechanisms.

Finally, it is well known that context-aware security can be used either for authentication, authorization and access control, and privacy-preserving [13]. However, in the context of all the presented solutions for IoT security in this section, context-aware security stands out in providing privacy preservation. It enables the applications executing on VMs to be able to set dynamic rules for privacy policies. Therefore the actions of providing/generating data to third-party applications are conditioned to the environmental context (i.e., characteristics of the deployed site). The reasoning module acts alongside with the rules to perform the privacy decisions when the environmental context changes (Fig. 9.11-J). This processing enables sensitive information to be shared/delivered to applications only with the accordance of the device owner. The owner can set the rules defining when to share data, how widely it will be shared, and whether some modification needs to be made to ensure data privacy. Context-aware security can avoid passive and active attacks once it protects the IoT device data in sensitive environments accordingly to the context. This prevents any unauthorized access or modification to such data. Again, the context-aware security code can be implemented as a VM, thus taking advantage of isolation and the TEE.

9.7.3 An application sample

Security in edge devices is an essential step towards protecting IoT applications. Let us assume the following scenario: An edge device, such as a heart rate monitor (HRM) is part of a home care system. It enables healthcare providers to remotely monitor and evaluate the health of patients in the comfort of their homes. This device has an HRM application, which should notify a hospital when a significant event occurs.

The proposed architecture should protect the integrity of the patient's sensitive data and protect against potential attacks that could compromise the device. Therefore the HRM application can be built as four different applications: (1) sensor monitoring, (2) decision-making, (3) context-aware security, and (4) fog/cloud communication. The sensor monitoring application is responsible for communicating with the data-acquisition devices. The inputs are transferred to the decision-making application through secure inter-VM communication. The decision-making application implements rules to perform data analysis of the patient's vital signs. Again, using secure inter-VM communication, the fog/cloud communication application uses the blockchain trusted network to send alerts to a cloud server if anomalies are detected. Finally, the context-aware security monitors all system activities, guaranteeing the user's data privacy.

9.8 Literature review

This section presents a literature review and a comparison of the proposed model with existing solutions. There are some works exploring TEEs and virtualization. Pinto et al. [57] propose a TrustZone-based architecture named IIoTTEED, which implements a TEE to protect edge devices. They concluded that IIoTEED should be complemented with other security strategies to guarantee tight industrial security for devices. Jang et al. [58] propose SeCRet, a framework that builds secure communications between a rich execution environment (REE) and TEE. SeCReT creates a session key to sign the messages transferred during interdomain communication. Results show that enabling SeCReT creates a performance overhead of 16.41% (from 1642.5 to 1912.1 μs) with an input payload of 256 B.

In our approach, the total footprint (SRAM and code storage) including TEE and security by separation mechanisms requires just 123 KB, which illustrates the small footprint of our architecture. Also, secure inter-VM communications happen through the use of hypercalls (99.59 μs for 256 B). The hypervisor is responsible for orchestrating messages from one VM to another. For comparison purposes, off-the-self hypervisors, such as Xen and KVM [29], require tens of megabytes. Xvisor [35], a hypervisor designed for embedded systems, requires up to 16 MB of RAM.

Regarding blockchain and context-aware security, there are a lot of research toward the protection of IoT systems and devices. However, there is a lack of research trying to integrate them as a way of strengthening the security of IoT devices, as proposed in our work. Rifi et al. [59] present an architecture based on a publish/subscribe model and use blockchain and smart contracts to secure IoT systems. Polyzos and Fotiou [60] use blockchain and smart contracts to improve trust in IoT environments. Ouechtati et al. [61] propose the development of a self-adaptive middleware to improve the access control of IoT applications based on a context-aware model. Hosseinzadeh et al. [54] propose a CARBAC scheme. The scheme controls the access of the users to a system following their role and the current context information. Trnka and Cerny [52] propose a solution that extends RBAC with context awareness elements. It is based on using security levels, which are granted to a user based on his context. Zhang et al. [62] designed a context-aware privacy preserving algorithm (CAPP) for smartphone applications to decide in which way the context of a user can be disseminated.

Finally, it is important to consider cloud and fog/edge computing paradigms in this discussion. According to the OpenFog Reference Architecture [5], most of computing, communications, and control is migrating to the edge of the network. Hence new fog and edge computing architectures are being defined [63, 64]. Morabito et al. [65] and Markakis et al. [66] explore the use of virtualization and containerization for edge devices, respectively. Also, Pattnaik et al. [67] discuss the progression in computing, analytics, and networking, while Mishra et al. [68] explore computational intelligence in sensor networks, which are relevant contributions to the fog/edge area. In our work, we defined a comprehensive security architecture for IoT devices, in which the main contributions are its low overhead and small footprint design.

However, it is important to reiterate that the cloud will still exist in the architecture and will have an important role in connecting devices [3]. Recent works discuss data scheduling [69], resource allocation [70], security [71], and nature-inspired optimizations [72], which are open research issues for cloud and fog/edge computing.

9.9 Conclusion

As we have seen, fog and edge computing paradigms are vital to the consolidation of the IoT. However, the security breaches in private networks by IoT edge devices are a major concern to IoT's broad adoption. In this chapter, we discussed four different technologies capable of improving the overall security of devices. These technologies cover different kinds of vulnerabilities, requiring their integration and cooperation in the same machine. As a result, we presented a security architecture flexible enough to allow for such combination. The main limitation of our approach is the overhead imposed by cryptographic operations required by the CoT and the blockchain. Even so, many embedded processors already add specialized hardware for cryptography and there are many small cryptographic libraries for tiny devices. In future work, we plan to explore techniques to decrease the protection's footprint and overhead, paving the way for the adoption of our architecture for deeply embedded systems.

References

[1] S. Kraijak, P. Tuwanut, A survey on IoT architectures, protocols, applications, security, privacy, real-world implementation and future trends, in: 11th WiCOM 2015, 2015, pp. 1–6.

[2] R. Mahmoud, T. Yousuf, F. Aloul, I. Zualkernan, Internet of things (IoT) security: current status, challenges and prospective measures, in: 10th International Conference for Internet Technology and Secured Transactions (ICITST), 2015, pp. 336–341.

[3] B.S.P. Mishra, H. Das, S. Dehuri, A.K. Jagadev, Cloud Computing for Optimization: Foundations, Applications, and Challenges, Springer, New York, 2018.

[4] M. Chiang, T. Zhang, Fog and IoT: an overview of research opportunities, IEEE Internet Things J. 3 (6) (2016) 854–864.

[5] OpenFog Consortium, OpenFog reference architecture for Fog computing, Tech. Rep. (02), 2017.

[6] P. Zhang, M. Zhou, G. Fortino, Security and trust issues in Fog computing: a survey, Futur. Gener. Comput. Syst. 88 (2018).

[7] A. Alrawais, A. Alhothaily, C. Hu, X. Cheng, Fog computing for the internet of things: security and privacy issues, IEEE Internet Comput. 21 (2017) 34–42.

[8] C. Moratelli, S. Johann, M. Neves, F. Hessel, Embedded virtualization for the design of secure IoT applications, in: Int. Symposium on Rapid System Prototyping (RSP), 2016, pp. 1–5.

[9] K. Sandstrom, A. Vulgarakis, M. Lindgren, T. Nolte, Virtualization technologies in embedded real-time systems, in: 18th Conference on Emerging Technologies Factory Automation (ETFA), 2013, pp. 1–8.

[10] K. Christidis, M. Devetsikiotis, Blockchains and smart contracts for the internet of things, IEEE Access 4 (2016) 2292–2303.

[11] M. Conoscenti, A. Vetr, J.C.D. Martin, Blockchain for the internet of things: a systematic literature review, in: Int. Conf. of Computer Systems and Applications (AICCSA), 2016, pp. 1–6.

[12] E. de Matos, L.A. Amaral, F. Hessel, Context-Aware Systems: Technologies and Challenges in Internet of Everything Environments, Springer International Publishing, Cham, 2017, pp. 1–25.

[13] E. de Matos, R.T. Tiburski, L.A. Amaral, F. Hessel, Providing context-aware security for IoT environments through context sharing feature, in: 17th International Conference on Trust, Security and Privacy in Computing and Communications (TrustCom), 2018, pp. 1711–1715.

[14] D.M. Mena, I. Papapanagiotou, B. Yang, Internet of things: survey on security, ISJGP 27 (3) (2018) 162–182.

[15] M. Sabt, M. Achemlal, A. Bouabdallah, Trusted execution environment: what it is, and what it is not, in: IEEE Trustcom, vol. 1, 2015, pp. 57–64.

[16] A. Greenberg, Hackers Remotely Kill a Jeep on the HighwayWith Me in It, https://www.wired.com/2015/07/hackers-remotely-kill-jeep-highway/. Accessed 20 September 2018.

[17] A. Greenberg, The Jeep Hackers Are Back to Prove Car Hacking Can Get Much Worse, https://www.wired.com/2016/08/jeep-hackers-return-high-speed-steering-acceleration-hacks/. Accessed 20 September 2018.

[18] E. Fernandes, J. Jung, A. Prakash, Security analysis of emerging smart home applications, in: Symposium on Security and Privacy, 2016, pp. 636–654.

[19] M. Antonakakis, et al., Understanding the Mirai Botnet, in: 26th USENIX Conference on Security Symposium, SEC'17, USENIX Association, Berkeley, CA, 2017, pp. 1093–1110.

[20] S. Mirzamohammadi, A.A. Sani, The case for a virtualization-based trusted execution environment in mobile devices, in: Asia-Pacific Workshop on Systems, NY, 2018, pp. 1–8.

[21] C. Herder, M. Yu, F. Koushanfar, S. Devadas, Physical unclonable functions and applications: a tutorial, Proc. IEEE 102 (8) (2014) 1126–1141.

[22] Z. Zhang, M.C.Y. Cho, C. Wang, C. Hsu, C. Chen, S. Shieh, IoT security: ongoing challenges and research opportunities, in: 7th Int. Conf. on Service-Oriented Computing and Applications, 2014, pp. 230–234.

[23] L. Atzori, A. Iera, G. Morabito, The internet of things: a survey, Comput. Netw. 54 (15) (2010) 2787–2805.

[24] K. Cheng, Y. Bai, R. Wang, Y. Ma, Optimizing soft real-time scheduling performance for virtual machines with SRT-Xen, in: 15th International Symposium on Cluster, Cloud and Grid Computing (CCGrid), 2015, pp. 169–178.

[25] J. Smith, R. Nair, Virtual Machines: Versatile Platforms for Systems and Processes (The Morgan Kaufmann Series in Computer Architecture and Design), Morgan Kaufmann Publishers, San Francisco, CA, 2005.

[26] F. Bellard, QEMU, a fast and portable dynamic translator, in: Annual Conference on USENIX Annual Technical Conference, Berkeley, CA, 2005, p. 41.

[27] P. Barham, B. Dragovic, K. Fraser, S. Hand, T. Harris, A. Ho, R. Neugebauer, I. Pratt, A. Warfield, Xen and the art of virtualization, SIGOPS Oper. Syst. Rev. 37 (5) (2003) 164–177.
[28] I. Habib, Virtualization with KVM, Linux J. (166) (2008). https://dl.acm.org/citation.cfm?id=1344217&preflayout=tabs.
[29] C. Dall, J. Nieh, KVM/ARM: The Design and Implementation of the Linux ARM Hypervisor, in: 19th International Conference on Architectural Support for Programming Languages and Operating Systems, New York, NY, 2014, pp. 333–348.
[30] S. Yoo, C. Yoo, Real-time scheduling for xen-arm virtual machines, IEEE Trans. Mob. Comput. 13 (8) (2014) 1857–1867, https://doi.org/10.1109/TMC.2013.109.
[31] A. Avanzini, P. Valente, D. Faggioli, P. Gai, Integrating Linux and the real-time ERIKA OS through the Xen hypervisor, in: Int. Symp. Industrial Embedded Systems (SIES), 2015, pp. 1–7.
[32] W. Kanda, Y. Yumura, Y. Kinebuchi, K. Makijima, T. Nakajima, SPUMONE: lightweight CPU virtualization layer for embedded systems, in: 2008 International Conference on Embedded and Ubiquitous Computing, vol. 1, 2008, pp. 144–151.
[33] S. Trujillo, A. Crespo, A. Alonso, MultiPARTES: multicore virtualization for mixed-criticality systems, in: Euromicro Conference on Digital System Design (DSD), 2013, pp. 260–265.
[34] R. Zhou, Z. Ai, J. Yang, Y. Chen, J. Li, Q. Zhou, K.-C. Li, A hypervisor for MIPS-based architecture processors—a case study in Loongson processors, in: International Conference on High Performance Computing and Communications, 2013, pp. 865–872.
[35] A. Patel, M. Daftedar, M. Shalan, M. Watheq El-Kharashi, Embedded Hypervisor Xvisor: A Comparative Analysis, in: 23rd Euromicro International Conference on Parallel, Distributed and Network-Based Processing (PDP), 2015, pp. 682–691.
[36] T. Xia, J.-C. Prevotet, F. Nouvel, Mini-NOVA: a lightweight ARM-based virtualization microkernel supporting dynamic partial reconfiguration, in: International Parallel and Distributed Processing Symposium Workshop (IPDPSW), 2015, pp. 71–80.
[37] G. Heiser, B. Leslie, The OKL4 microvisor: convergence point of microkernels and hypervisors, in: Asia-Pacific Workshop on Workshop on Systems, NY, 2010, pp. 19–24.
[38] G.J. Popek, R.P. Goldberg, Formal requirements for virtualizable third generation architectures, Commun. ACM 17 (7) (1974) 412–421.
[39] N. Kshetri, Can Blockchain strengthen the internet of things? IT Prof. 19 (4) (2017) 68–72.
[40] D. Johnson, A. Menezes, S. Vanstone, The elliptic curve digital signature algorithm (ECDSA), Int. J. Inf. Secur. 1 (1) (2001) 36–63.
[41] H. Gilbert, H. Handschuh, Security analysis of SHA-256 and sisters, in: M. Matsui, R. J. Zuccherato (Eds.), Selected Areas in Cryptography, Springer, Berlin; Heidelberg, 2004, pp. 175–193.
[42] H. Wang, Z. Zheng, S. Xie, H.-N. Dai, X. Chen, Blockchain challenges and opportunities: a survey, Int. J. Web Grid Serv. 14 (2018) 352.
[43] A. Banafa, IoT and Blockchain convergence: benefits and challenges, IEEE Internet Things (2017).
[44] J. Moubarak, E. Filiol, M. Chamoun, On blockchain security and relevant attacks, in: Middle East and North Africa Communications Conference (MENACOMM), 2018, pp. 1–6.
[45] A. Dorri, S.S. Kanhere, R. Jurdak, P. Gauravaram, Blockchain for IoT security and privacy: the case study of a smart home, in: International Conference on Pervasive Computing and Communications Workshops (PerCom Workshops), 2017, pp. 618–623.

[46] A. Dorri, S.S. Kanhere, R. Jurdak, Towards an Optimized BlockChain for IoT, in: 2nd Int. Conf. on Internet-of-Things Design and Implementation, New York, NY, 2017, pp. 173–178.
[47] P. Brezillon, G.K. Mostefaoui, Context-based security policies: a new modeling approach, in: Conference on Pervasive Computing and Communications Workshops, 2004, pp. 154–158.
[48] C. Perera, A. Zaslavsky, P. Christen, D. Georgakopoulos, Context aware computing for the internet of things: a survey, IEEE Commun. Surv. Tutorials 16 (1) (2014) 414–454.
[49] G.D. Abowd, A.K. Dey, P.J. Brown, N. Davies, M. Smith, P. Steggles, Towards a better understanding of context and context-awareness, in: 1st International Symposium on Handheld and Ubiquitous Computing, Springer-Verlag, London, 1999, , pp. 304–307.
[50] J.L.H. Ramos, J.B. Bernabe, A.F. Skarmeta, Managing context information for adaptive security in IoT environments, in: 29th International Conference on Advanced Information Networking and Applications Workshops, 2015, pp. 676–681.
[51] K. Habib, W. Leister, Context-aware authentication for the internet of things, in: Eleventh International Conference on Autonomic and Autonomous Systems, 2015, p. 6.
[52] M. Trnka, T. Cerny, On security level usage in context-aware role-based access control, in: 31st Symposium on Applied Computing, New York, NY, 2016, pp. 1192–1195.
[53] E. Hayashi, S. Das, S. Amini, J. Hong, I. Oakley, CASA: context-aware scalable authentication, in: Ninth Symposium on Usable Privacy and Security, New York, NY, 2013, pp. 3:1–3:10.
[54] S. Hosseinzadeh, S. Virtanen, N. Díaz-Rodríguez, J. Lilius, A semantic security framework and context-aware role-based access control ontology for smart spaces, in: International Workshop on Semantic Big Data, New York, NY, 2016, pp. 8:1–8:6.
[55] M. Langheinrich, A Privacy Awareness System for Ubiquitous Computing Environ ments, Springer, Berlin, Heidelberg, 2002, pp. 237–245 (Chapter 19).
[56] A. Pingley, W. Yu, N. Zhang, X. Fu, W. Zhao, A context-aware scheme for privacy-preserving location-based services, Comput. Netw. 56 (11) (2012) 2551–2568.
[57] S. Pinto, et al., IIoTEED: an enhanced, trusted execution environment for industrial IoT edge devices, IEEE Internet. Comput. 21 (1) (2017) 40–47.
[58] J.S. Jang, et al., SeCReT: secure channel between rich execution environment and trusted execution environment, in: Network and Distributed System Security Symposium (NDSS), 2015, pp. 1–15.
[59] N. Rifi, E. Rachkidi, N. Agoulmine, N.C. Taher, Towards using blockchain technology for IoT data access protection, in: 17th International Conference on Ubiquitous Wireless Broadband (ICUWB), 2017, pp. 1–5.
[60] G.C. Polyzos, N. Fotiou, Blockchain-assisted information distribution for the internet of things, in: International Conference on Information Reuse and Integration (IRI), 2017, pp. 75–78.
[61] H. Ouechtati, N.B. Azzouna, L.B. Said, Towards a self-adaptive access control middleware for the Internet of Things, in: Int. Conf. on Information Networking (ICOIN), 2018, pp. 545–550.
[62] L. Zhang, Y. Li, L. Wang, J. Lu, P. Li, X. Wang, An efficient context-aware privacy preserving approach for smartphones. Secur. Commun. Netw. 2017 (2017) 1–11, https://doi.org/10.1155/2017/4842694.
[63] A. Munir, P. Kansakar, S.U. Khan, IFCIoT: integrated Fog cloud IoT: a novel architectural paradigm for the future internet of things, Consum Electr. Mag. 6 (3) (2017) 74–82.
[64] Y. Sahni, J. Cao, S. Zhang, L. Yang, Edge mesh: a new paradigm to enable distributed intelligence in internet of things, IEEE Access 5 (2017) 16441–16458.

[65] R. Morabito, R. Petrolo, V. Loscr'ı, N. Mitton, Enabling a lightweight Edge Gateway-as-a-Service for the Internet of Things, in: 7th International Conference on the Network of the Future (NOF), 2016, pp. 1–5.
[66] E.K. Markakis, K. Karras, N. Zotos, A. Sideris, T. Moysiadis, A. Corsaro, G. Alexiou, C. Skianis, G. Mastorakis, C.X. Mavromoustakis, E. Pallis, EXEGESIS: extreme edge resource harvesting for a virtualized fog environment, IEEE Commun. Mag. 55 (7) (2017) 173–179.
[67] P.K. Pattnaik, S.S. Rautaray, H. Das, J. Nayak, Progress in Computing, Analytics and Networking: Proceedings of ICCAN 2017, vol. 710, Springer, 2018.
[68] B.B. Mishra, S. Dehuri, B.K. Panigrahi, A.K. Nayak, B.S.P. Mishra, H. Das, Computational Intelligence in Sensor Networks, Springer, New York, 2019.
[69] K.H.K. Reddy, H. Das, D.S. Roy, A data aware scheme for scheduling big data applications with savanna hadoop, in: Networks of the Future, Chapman and Hall/CRC, New York, 2017, pp. 377–392.
[70] H. Das, A.K. Jena, J.C. Badajena, C. Pradhan, R. Barik, Resource allocation in cooperative cloud environments, in: Progress in Computing, Analytics and Networking, Springer, New York, 2018, pp. 825–841.
[71] C. Pradhan, C. Pradhan, H. Das, B. Naik, N. Dey, Handbook of Research on Information Security in Biomedical Signal Processing, 1st ed., IGI Global, Hershey, PA, 2018.
[72] J. Nayak, B. Naik, A. Jena, R.K. Barik, H. Das, Nature inspired optimizations in cloud computing: applications and challenges, in: Cloud Computing for Optimization: Foundations, Applications, and Challenges, Springer, 2018, pp. 1–26.

CHAPTER

Software-Defined Networking for the Internet of Things: Securing home networks using SDN

10

Shivaranjani Anbarsu, Arockia Xavier Annie Rayan, Vettriselvi Vetrian
Department of Computer Science and Engineering, Anna University, Chennai, India

10.1 Introduction

A Software-Defined Network (SDN) is helpful in connecting smart digital appliances with members of the family in our houses today. Diversity and heterogeneity of these connected home devices are handled easily using SDN. This is because the appliances introduce protocols with various management and interoperability challenges [1]. For example, multimedia appliances have different processing powers, screen resolutions, network connection speeds, and home-networking protocols. To overcome these obstacles, various solutions have been proposed, including the provisioning of intra-home quality of service (QoS) as proposed by [2], interoperability with UniDA, and network address translation (NAT) traversal as proposed by Yoon et al [3]. However, these solutions do not cover the requirements for flexible data sharing because of prearranged networking protocols. This causes fragmentation as a major challenge in home networks. This is resolved using SDN, as it helps in configuring home devices through its control plane. It also facilitates flexible data sharing among home devices.

With SDN-enabled home platforms, device configurations and packet control are performed with ease. Otherwise these tasks are time consuming, cumbersome, and done manually. In order to configure multiple devices and to enable them to interoperate, major technical issues such as scalability, consistency, security, and privacy need to be addressed.

The interoperability issues between home consumer electronic devices in a single network were addressed by Yoon et al [3]. They proposed a tunnel-based peer-to-peer (p2p) communication scheme that guaranteed seamless and secure data exchange between mobile Internet devices behind NAT [3]. Client-/server-based architecture was used for seamless and secure p2p communication. The client module was mounted on the mobile nodes (MNs) and the server module managed the

Real-Time Data Analytics for Large Scale Sensor Data. https://doi.org/10.1016/B978-0-12-818014-3.00010-3

clients in real-time using the unique identifiers assigned to each client. Interoperability and management between multiple home networks were not addressed.

Boussard et al. proposed a software-based solution called SD-LAN to interconnect smart devices according to their services. In this system, SDN controls and manages parts of a smart environment through their virtual representation, which also interconnect by using SDN's architectural principles [4]. Here a home network is interconnected temporarily to the existing multihomed network. It also provided user-level privileges in accessing the network. Issues such as SD-LAN scalability, privacy, and security were not addressed.

In a smart home system, a community connector is used for integrating community services, thereby reducing the workload of the management routine [5]. A Message Queuing Telemetry Transport (MQTT) protocol was used in smart home systems to provide control services, while HTTP was used to deliver location-based information integration services [6]. The proposed solutions were used to provide various services based on the IoT context. It solely concentrated on how to integrate IoT devices in a multihomed environment to provide various services. However, issues with scalability, privacy, and security in the multihomed environment were not addressed.

Jo et al. redesigned a home router [1], and its prototype was implemented using NOX and OpenFlow to provide per-flow control. A custom Dynamic Host Configuration Protocol (DHCP) was also implemented to enable traffic isolation and accurate measurement from the IP layer [7]. The home router leveraged SDN technology to provide home users with control over the behavior of its network [7]. The user had control over their single-home network behavior but not in a multihomed network environment.

A better way to deal with security issues in home networks is to outsource the management and operation of networks to a common node that has both operations management expertise and complete view of network activity within the networks limit. As proposed by Feamster et al. [8], an offsite controller detects Internet-wide coordinated activities arising from home networks that automates necessary actions through controllable network gateways for the sake of the home user. The user need not have to worry about security issues because the third-party controller takes care of it. Appropriate balance between user privacy and security will be difficult.

Smart home networks use cloud computing technology to aid decision making. As proposed by Tom et al. [9], a risk-based integrated management of devices was followed to improve the utilization of home resources. Using automated risk assessments linked to a service-level agreement (SLA) and user/application requirements the smart home can get smarter. Cloud computing helped in managing the resources in a home network. Data privacy and manifested contracts of data consumption via sharing and usage are needed in the form of SLA and user notifications [10, 11].

One of the major issues in OpenFlow switch is its limited-size flow table, which results in displacement of flows from the flow table [8]. This work uses a hash-based placement and localized Least Recently Used (LRU)-based replacement mechanisms [12]. The algorithm is invoked only when the flow table becomes full.

The work described by Kreutz et al. [13] defines various issues related to scalability. It addresses switch design and control platforms that focus on resiliency, scalability, performance, security, and dependability. It also paved the way for newer opportunities for carrier transport networks and cloud providers. They also purview the position of SDN as a sole descriptor of the software-defined environment. Numerous SDN solutions were proposed that allowed home users to have control over their networks' behavior.

Next, addressing the security issues in home networks, even with minimal false alarm rates, the IDS must be exceedingly capable of achieving a high attack-detection rate. To deliver accurate IDS, machine-learning and data-mining techniques, integrated with computational intelligence, are studied and tried. The use of Genetic Fuzzy Systems within a pair-wise learning were used by Elhag et al. [14] for implementing optimal IDS. This increased the benefits in two ways. The first being a fine and clear separation between the concepts, and providing a higher interpretability of the rule set through the use of fuzzy sets with linguistic labels. The second bein the divide-and-conquer learning scheme that demarcated pairs of classes that were clear contrast. This greatly improved the precision for the rare attack events, as it obtained a better separation between a "normal activity" and the different attack types. Here, a KDD CUP99 dataset was used for evaluation and testing.

Another optimal IDS using an Adaptive Neuro-Fuzzy Inference System (ANFIS) was studied and implemented [15] by Devi et al. on 5G wireless networks using a KDD CUP99 dataset for detecting an attack on the relay. The effect of modifying the membership function and machine learning algorithms are the major areas of study and analysis for accurate IDS. A device-to-device communication scenario was considered in implementing the Fuzzy Inference System (FIS), in which the end user uses relay to receive the signal. With relay in the communication network, numerous attacks have been attempted, including Masquerade, Repudiation, and Denial-of-Service (DoS) attacks, among others. These attacks are achieved by the intruder in the network attacking the relay and acquiring the user's information connected to the relay. In this work, a KDD CUP99 dataset was used for training the generated FIS model that calculates threshold value of average testing error. Among the different types of attacks in the KDD CUP99 dataset, only DoS-based attacks were considered for analyzing the proposed model.

A Bagging Ensemble method was used [16] by implementing an IDS. REPTree was used as the base class. The required feature set from NSL_KDD dataset was selected to improve classification accuracy. The experimental results showed that the Bagging Ensemble with REPTree base class achieved the highest classification accuracy. The Bagging Ensemble method took less time in building the model and provided low false positive rates compared with other machine learning techniques such as C4.5, random forest, and naïve Bayes.

Deep learning technique was applied, for flow-based anomaly detection in SDN, for this an IDS environment was applied by Tuan et al. [17]. A Deep Neural Network (DNN) model was built and trained with the NSL-KDD dataset. Only six basic features (that can be easily obtained in an SDN environment) were taken from the

41 features of the NSL-KDD dataset in training the model. The DNN model performed better than other machine learning algorithms such as J48, naïve Bayes, multilayer perceptron, random forest, and random tree [18–20].

Deep learning approaches were used in implementing a botnet IDS. Cyber security events have created awareness in harnessing the need for machine learning algorithms that adapt to evolving network systems. Signature-based approaches for identifying data types were useless in finding the attacks that took place on normal attack platforms. Human intervention proved unfit with encrypted and altered data. A neural network model can learn to identify subtle patterns in a suitably chosen input space [21–24]. A signal processing approach was introduced for classifying data files, which instantly changes to new data formats. DNN was evaluated for three input spaces consisting of the power spectral density, byte probability distribution, and sliding window entropy of the byte sequence in a file [23]. By combining all three the trained DNN was able to discriminate amongst nine common data attacks found on the Internet with 97.4% accuracy.

In a vehicular network, the deep learning approach was used in implementing an IDS. The nodes in the DNN structure are built with probability-based feature vectors from the in-vehicular network packets [25]. As compared to the traditional ANNs applied to the IDS, this technique adopts recent advances in deep learning studies, such as initializing the parameters through the unsupervised pretraining of DNNs therefore improving detection accuracy.

10.2 Methodology

10.2.1 Scalability and consistency

Integrating multiple home networks is a challenging task due to incompatible networking protocols and heterogeneous resources. The scalability property allows the existing home network to add or remove SDN-enabled devices on demand. It also allows for adding or removing a new home network. When one device is sharing its data to two other devices located at different home networks, then the data must be consistent in the other two home networks. This problem of inconsistency should be avoided. Various issues that arise with scalability are discussed in the following paragraphs.

(i) Flow-based forwarding

There are two types of flow-based forwarding, namely, proactive flow forwarding and reactive flow forwarding. Proactive flow forwarding is similar to traditional routing. The network topology is known beforehand and forwarding entries are installed. Reactive flow forwarding refers to the punting of unknown packets to the controller and the installation of flow entries based on actual traffic with no prior installation required.

a. Scalability concerns in flow-based forwarding

Packet Punting Rate: This refers to how many packets the flow can send to a controller. This depends on the switch CPU and the bandwidth between the forwarding fabric and the switch CPU.

Flow Modification Rate: This refers to how many flows per second a controller can push down to the switches. This is a major issue because the flow table size is limited. Nowadays, production switches can handle only a thousand notifications per second.

b. Path-based forwarding vs. hop-by-hop forwarding (proactive)
Hop-by-hop forwarding is similar to classic IP forwarding wherein forwarding is based on hop count. Path-based forwarding is similar to multiprotocol label switch (MPLS) wherein the path is computed across the network. The map flows to the best path.

(ii) Scalability concerns in path-based forwarding vs. hop-by-hop forwarding

- number of flow updates after topology change
- limited fast re-route capability

Path-based forwarding is better than hop-by-hop forwarding because of the reduced number of core flows and easy implementation of fast rerouting.

(iii) Central controller
Handling of requests by the central controller becomes a bottleneck as far as scalability is concerned. The question of how many multiple controllers are required so that we can scale in and scale out is a difficult one to answer.

10.2.2 Security and privacy

The system sends raw packets to the controller that needs to be encrypted before being sent to the controller. When adding a new home network, authentication and authorization must take place. Multihomed networks are susceptible to malicious attacks by intruders. All these issues pose a security threat to the multihomed network environment. In a SDN network, the separation of control plane from the data plane provides several advantages. The centralized controller can monitor the packets by deciding whether a packet can be blocked or allowed to be sent to the data plane. The controller can also set priorities, set up/delete flow in the flow table, collect network statistics, and set up the path for the packet flow. The OpenFlow protocol is the protocol used by the SDN controllers to communicate with the switches/routers.

Since the controller is centralized, it is susceptible to several security issues just like conventional networks. In particular, SDN environments [26] can be easily attacked by DoS attacks, man-in-the-middle attacks, and probe attacks. DoS attacks can degrade the performance of SDN networks by denying services to legitimate requests. Man-in-the-middle attacks can disrupt the communication between the controller and networking devices [23]. Probe attacks can gather sufficient information about the network to find vulnerabilities that can be exploited to send a future attack. Hence security mechanisms should be built on top of the controller to deal with these malicious attacks by intruders. An IDS that can detect these malicious activities is essential in such types of environments.

There are various techniques used to implement IDS [15, 25, 27, 28]. Traditional IDS made use of rules from rule-based expert systems to detect any known pattern of attack. The major disadvantage of this technique is that the rules need to be manually installed. In addition, it does not offer much flexibility in an evolving network. In order to overcome this issue, machine-learning techniques were employed in implementing an IDS [28, 29]. A supervised learning approach is used in the detection of malicious attacks since it is a classification problem.

10.3 System design

The proposed system aims at providing an optimized solution for decreasing the flow modification rate using an "optimized flow replacement manager." This is based on "Proactive Rule Replacement," where replacement takes place even before the flow table becomes full. The replacement follows LRU policy as given in Eq. (10.1).

The module is invoked when,

$$P >= (3/4)*N \tag{10.1}$$

where P is the number of flow entries in the table and N is the total capacity of the flow table.

After this, the packet is forwarded to the flow replacement manager. Later, it looks up the flow buffer entries for the respective switch. The flow entry with maximum duration and minimum counter value is chosen for replacement.

The environment consists of virtual machine_1 acting as homenetwork_1 and a virtual machine_2 acting as homenetwork_2. The ControlBox (control plane-SDN) is responsible for coordinating and to control the event activities between the multihomed network as shown in Fig. 10.1. The user first sends an Application Description (AD) message containing information such as type of service, IP addresses of gateway HomeBoxes, match rule of a flow, and so on. On receiving the request message, the ControlBox sends a control instruction consisting of configuration details to the two home networks. The Gateway HomeBox receives this control message and checks the flow table to find any match. If any match is found, then the appropriate action is taken.

10.3.1 Subsidiary HomeBox

This module consists of a user interface through which the user can send the AD message, add device message, and remove device message. We make use of Mininet (OpenFlow SDN simulator) to create hosts corresponding to various home networks. The user can send packets to other hosts in a different home network using a ping command through the Mininet command-line interface (CLI). The packets that a user sends will be forwarded to the switch. The switch receives the packet and checks its flowtable to find a match. If a match is not found, then by default it is forwarded to the controller. The controller then processes the packet and forwards necessary actions to the switch [29].

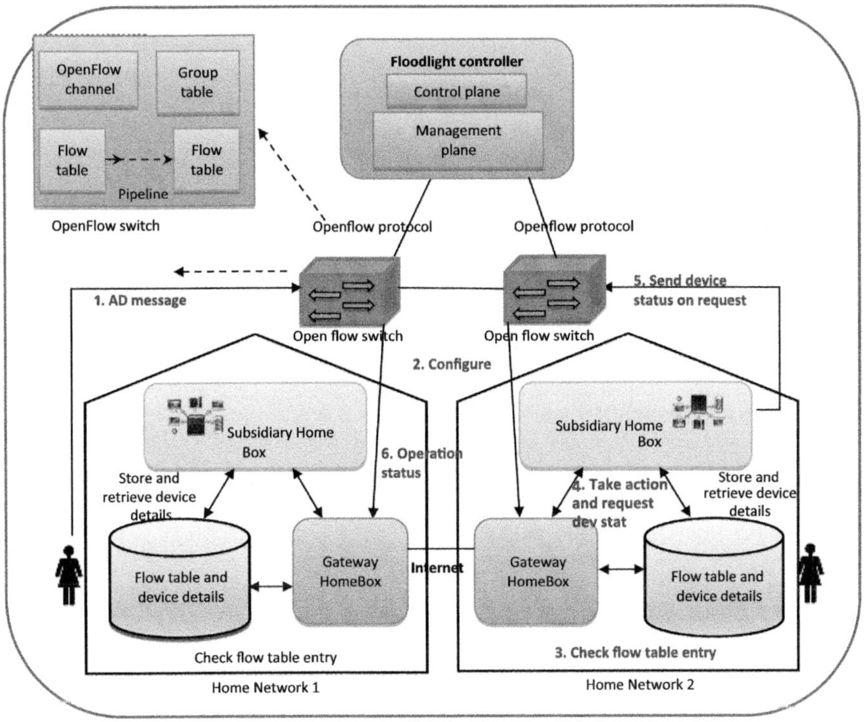

FIG. 10.1

Architecture diagram of a SDN-enabled multihomed networking environment with HomeBox and ControlBox.

- *User Interface (UI)*: functions such as send requests with (1) AD, (2) add device, and (3) remove device messages and displays device status with details.
- *Device Handler(DH)*: functions such as stores device details, device topology, device operations, device configurations, transfer data (audio, video, image, text files) from device to Gateway HomeBox for processing, send device topology and device details to controller, receive configuration instruction and change the configuration, receive data from Gateway HomeBox, which will be forwarded to respective user devices. The user interface setup is as shown in Fig. 10.2. The flow table in each switch consists of the following entries. Match rule is required to find matches in the entry which is as shown in Fig. 10.3. The match rules as shown in Fig. 10.4, typically consists of Ethernet destination MAC address, Ethernet source MAC address, ingress port, type of packet, source IP address, destination IP address, and metadata. The following diagram shows each field described in flow table entry.

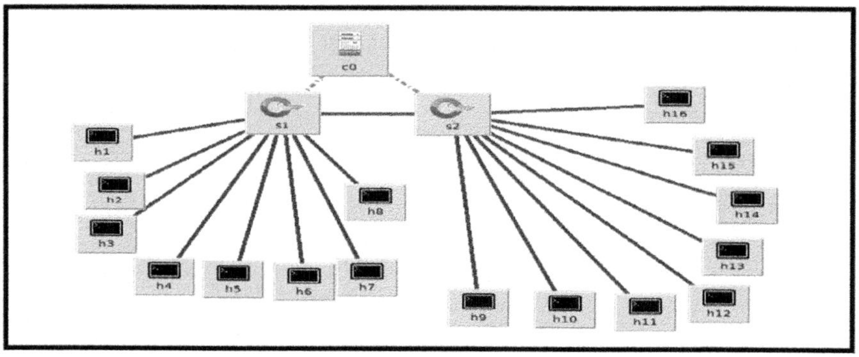

FIG. 10.2

User interface in Mininet SDN simulator.

FLOWTABLE ENTRY :

MAT_RULE	COUNTER	DURATION	IDLE TIMEOUT	HOP	PRIORITY	ACTION

CHARACTER		DESCRIPTION
MAT_RULE	=	Match rule of a flow
COUNTER	=	To update for matching packets
DURATION	=	Duration that a flow stays in the flow table
IDLE TIMEOUT	=	Idle time before flow is expired by the switch
HOP	=	Hop count
PRIORITY	=	Matching precedence of the flow entry
ACTION	=	Specifies action to be taken such as drop, forward and send to controller

FIG. 10.3

Flow table entry description.

Fields of Match rule (MAT-RULE):

SRC_IP	DEST_IP	PROTOCOL	SRC_PORT	DEST_PORT	SRC_ MAC	DEST_ MAC	META DATA	CTRL_ INSTR

CHARACTER		DESCRIPTION
SRC_IP	=	Source IP address
DEST_IP	=	Destination IP address
PROTOCOL	=	Type of protocol
SRC_PORT	=	Port address of source
DEST_PORT	=	Port address of destination
SRC_MAC	=	MAC address of source device
DEST_MAC	=	MAC address of destination device
METADATA	=	Metadata associated with control actions to be taken in home environment
CTRL_INST	=	Specifies the type of control messages from controller

FIG. 10.4

Fields of match rule and their description.

(i) AD Message Packet

The user sends an AD message to the switch. The message consists of information about the IP addresses of the Gateway HomeBox to which the home network is connected, the configuration details of the device, and so on. The switch forwards the packet to the controller if it is a new packet. On receiving the AD message, the controller searches the DEV_ID in its database. If DEV_ID is found, then it sends the appropriate configuration instruction to the respective Gateway HomeBoxes. If DEV_ID is not found, the controller drops the packet. Fig. 10.5 shows the flow diagram to perform the AD request message.

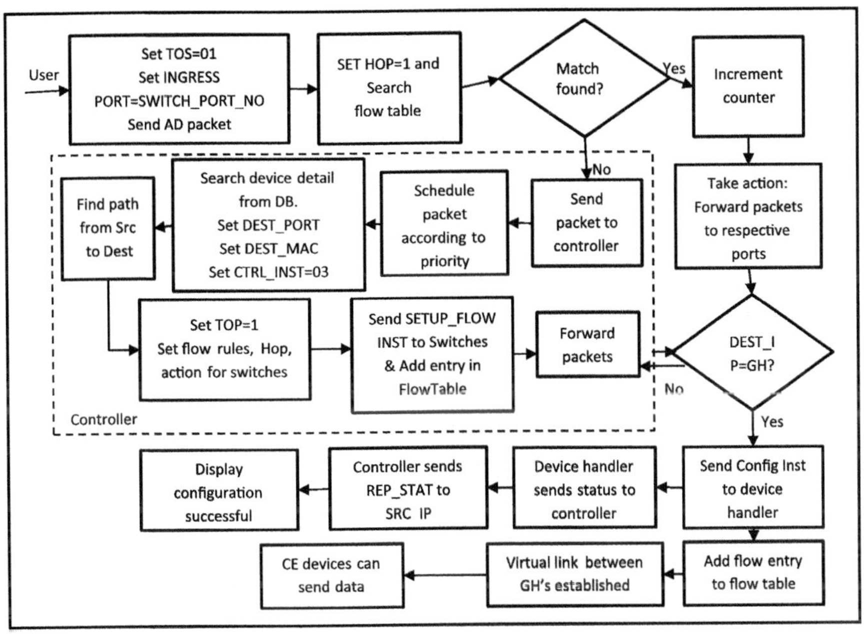

FIG. 10.5

Message flow when sending an AD request message.

INPUT:

```
AD Message Packet,
MATCH_RULE, Match Rule of a Flow,
HM_ADDR,IP addresses of Gateway HomeBox, TOS, Type of
  service, TOS=1,
PRIORITY, Defines the priority, INGRESS PORT, Port number of
  entry port,
HOP, Hop count
```

OUTPUT: `Display Configuration Successful`

Continued

ALGORITHM: ConfigDev()

```
 Set HOP=1
 If(FlowTable.Entry==Match) then
       increment Counter
 take action for forwarding
 else
     Forward AD to Controller
 queue the packets in QUEUE[],
 if(PACKET1.priority>PACKET2.priority) then
      send PACKET1 to packet processer
 else if(PACKET1. Priority=PACKET2.Priority)then
      if(PACKET1.Duration>PACKET2.Duration) then
             send PACKET2 to packet processor
 Search DEV_ID details in database
 set DEST_MAC in MAT_RULE
 set METADATA=CONFIG_SRCDEV_DESTDEV
 set DEST_PORT
 set CTRL_INST=03
 Send SRC_IP & DEST_IP to topology manager
 find the shortest Path between SRC_IP and DEST_IP
 for(switches.path){
      set HOP
      set ACTION
      set TOP=1;
  }Send SETUP_FLOW to switches.path
      ADD FLOWTABLE.entry
      Take actions for forwarding packet
 If reached DEST_IP then
  {     Send CONFIG_INST to Device Handler
        Device Handler sets CONFIG_SRCDEV_DESTDEV
               Device Handler sends the DEV_STAT to Controller
  }Contollersends REP_STAT to SRC_IP
 if(DEV_STAT==ON) then
  {  ADD FLOWTABLE.Entry
     Virtual link establishment successful
     devices can now send data
     display "CONFIGURATION SUCCESSFUL"
  }else
     display "DESTINATION DEVICE IS SWITCHED OFF"
End
```

(ii) ADD Device Message

The user sends an Add Device (ADDev) message to the switch. The message consists of information about the IP addresses of the Gateway HomeBox to which the home network is connected, the details of the device such as DEV_ID, DEV_STAT, and so on. The switch forwards the packet to the controller if it is a

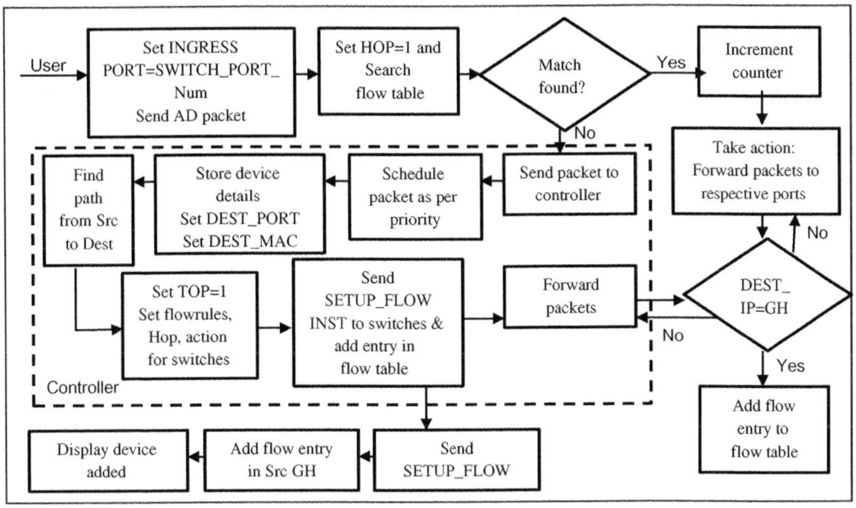

FIG. 10.6

Message flow when sending an add device message.

new packet, as shown in Fig. 10.6. On receiving the ADDev message, the controller stores the details of the device in its database and sends the appropriate control instructions to the switches and Gateway HomeBoxes.

INPUT:
```
ADDev message packet,
MATCH_RULE, Match Rule of a Flow,
HM_ADDR, IP addresses of Gateway HomeBox, TOS, Type of
 service, TOS=2,
PRIORITY, Defines the priority, INGRESS PORT, Port number of
 entry port,
HOP, Hop count, DEV_TOP, Details of device topology,
DEV_STAT, Device status, DEV_DET, DeviceID, Configuration and
 operations details.
```
OUTPUT: `Display "Device added Successfully"`

ALGORITHM: ADDevice()
```
Set HOP=1
If(FlowTable.Entry==Match) then
      increment Counter
take action for forwarding
else
     Forward ADDev to Controller
queue the packets in QUEUE[],
```

Continued

```
if(PACKET1.priority>PACKET2.priority) then
      send PACKET1 to packet processer
else if(PACKET1. Priority=PACKET2.Priority)then
      if(PACKET1.Duration>PACKET2.Duration) then
              send PACKET2 to packet processor
STORE DEV_ID in database
STORE DEV_DET in database
STORE DEV_TOP in database
Based on the DEV_OP performed by devices
find all possible combinations of configurations that can be done
set DEST_MAC in MAT_RULE
set METADATA=ADD_DEV
set DEST_PORT
set CTRL_INST=01
Send SRC_IP & DEST_IP to topology manager
find the shortest Path between SRC_IP and DEST_IP
for(switches.path)
{     set HOP
      set ACTION
      set TOP=1;
 }
Send SETUP_FLOW to switches.path
     ADD flowtable.entry
     Take actions for forwarding packet
If reached DEST_IP then
     ADD FLOWTABLE.entry to DEST_IP
contoller sends SETUP_FLOW to SRC_IP
If(CTRL_INST==1)
    display "DEVICE ADDED SUCCESSFULLY"
else If(CTRL_INST==*)
    display "ERROR ADDING DEVICE"
END
```

(iii) Remove Device Message

The user sends a REMDev message to the switch. The message consists of information about the IP addresses of the Gateway HomeBox to which the home network is connected, the details of the device such as DEV_ID, DEV_STAT, and so on. The switch forwards the packet to the controller if it is a new packet. On receiving the REMDev message, the controller deletes the details of the device in its database and sends the appropriate control instructions to the switches and Gateway HomeBoxes. The working module for REMDev is shown in Fig. 10.7.

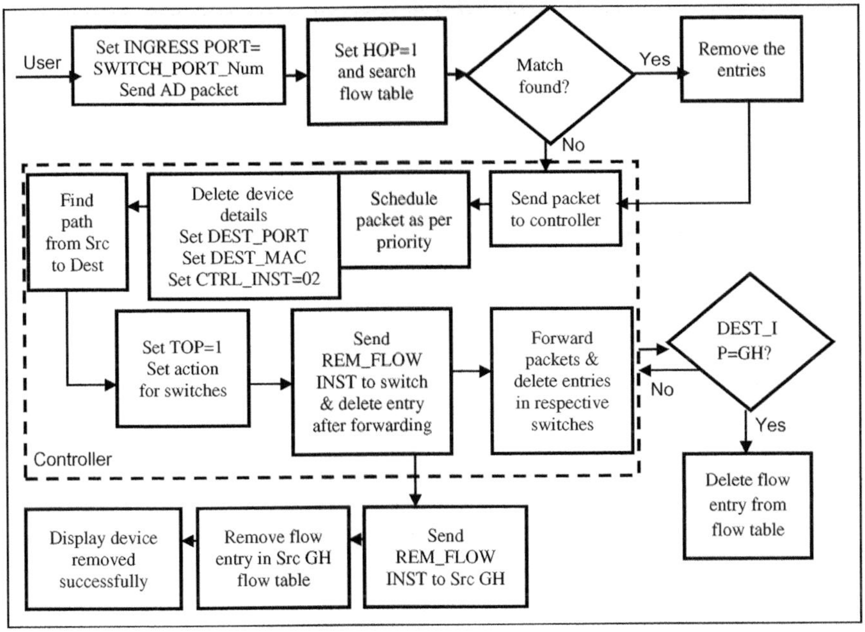

FIG. 10.7

Message flow when sending a remove device message.

INPUT:

```
     REMDev message packet, MAT_RULE, Match Rule of a Flow,
     HM_ADDR, IP addresses of Gateway HomeBox, TOS, Type of
     service, TOS=3, PRIORITY, Defines the priority,
     INGRESS PORT, Port number of entry port, HOP, Hop count,
     DEV_DET, DeviceID, Configuration and operations details.
```

OUTPUT: Display "removed device successfully"

ALGORITHM: REMDevice()

```
 Set HOP=1
 If(FlowTable.Entry==Match) then
      forwardREMDev to controller
       REM_FLOW entry
 else
       Forward REMDev to Controller
queue the packets in QUEUE[],
if(PACKET1.priority>PACKET2.priority) then
      send PACKET1 to packet processer
else if(PACKET1. Priority=PACKET2.Priority)then
      if(PACKET1.Duration>PACKET2.Duration) then
              send PACKET2 to packet processor
```

Continued

```
Remove DEV_ID details from database
set DEST_MAC in MAT_RULE
set METADATA=REM_DEV
set DEST_PORT
set CTRL_INST=02
Send SRC_IP & DEST_IP to topology manager
find the shortest Path between SRC_IP and DEST_IP
for(switches.path)
{
     forwardREMDev packet
      Remove FLOWTABLE.entry
      Set TOP=1;
}
Send REM_FLOW to switches.path
Remove FLOWTABLE.entry after forwarding packets
Take actions for forwarding packet
If reached DEST_IP then
      Remove FLOWTABLE.entry in DEST_IP flowtable
contoller sends REM_FLOW to SRC_IP
If(CTRL_INST==2)
    display "DEVICE REMOVED SUCCESSFULLY"
else If(CTRL_INST==*)
    display "ERROR REMOVING DEVICE"
END
```

10.3.2 Gateway HomeBox

The Gateway HomeBox is responsible for forwarding the control instructions from the controller to the devices in home networks. It also sends and receives data from various other Gateway HomeBoxes. Some of the responsibilities of the Gateway HomeBox include managing the flow table, processing the packet, and reporting information. The Gateway HomeBox does various operations based on the CTRL_INST field. It is also responsible for drefragmenting the data packets according to their sequence numbers. All the details are as shown in Fig. 10.8.

Resource Coordinator: queues the incoming packets and schedules them

Packet Processor: extracts the necessary details such as match rule, metadata from match rule, and performs defragmentation

Flow Table Manager: flow entry generation that performs flow modification and flow deletion along with match rule checking

Configuration Manager: sends configure instructions to the subsidiary HomeBox and keeps track of the list of devices that are configured

Information Reporter: sends the data packet to the device controller, device/op status to the UI, requests to send device status information to DH

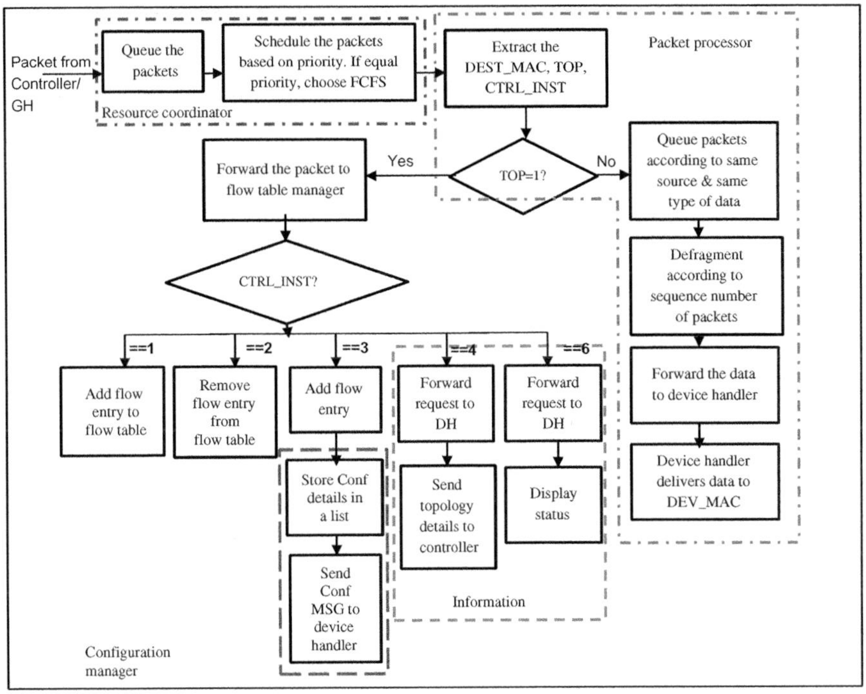

FIG. 10.8

Block diagram of Gateway HomeBox.

```
INPUT:
 CONTROL/DATA PACKET, SRC_IP, IP address of source, DEST_IP, IP
   address of destination, MAT_RULE, Match Rule of a Flow,
   TOP, Type of Packet, PRIORITY, Defines the priority,
   1-implies high priority, INGRESS PORT, Port number of entry
   port, HOP, Hop count
OUTPUT: Display "added/removed flow entry successfully"
ALGORITHM: GatewayHomeBox()
 queue the packets in QUEUE[],
 if(PACKET1.priority>PACKET2.priority) then
      send PACKET1 to packet manager
 else if(PACKET1. Priority=PACKET2.Priority)then
      if(PACKET1.Duration>PACKET2.Duration) then
             send PACKET2 to packet manager
 Extract DEST_MAC, TOP, CTRL_INST from Control/Data Packet
 Send the extracted value to FLowTable Manager
```

Continued

```
If(TOP==1)
{   if(CTRL_INST==01)
    {  ADD FLOWTABLE.entry
       Display "flow table entry added successfully"
    }else if(CTRL_INST==02)
    {  if(FLOWTABLE.entry==MATCH)
       {  REMOVE FLOWTABLE.entry
          Display "flow table entry removed successfully"
       }
    }else if(CTRL_INST==03)
    {  ADD FLOWTABLE.entry
       Configuration Manager sends CONFIG_INST to device handler
       device handler changes the device configuration.
    }else if(CTRL_INST==04)
    {  Information reporter sends a request to device handler
       device handler sends DEV_TOP information to controller
     }
     else if(CTRL_INST==05)
     {     Information Reporter sends a request to device handler
           device handler sends DEV_STAT information to controller
       }
     else if(CTRL_INST==06)
     {       Information Reporter sends the DEV_STAT to UI.
           Device status of DEST_IP will be displayed
       }
    } else if(TOP!=1)
    { queue the data packets to QUEUE1[] from same source and device;
      Perform defragmentation according to sequence number.
      forward the data to device handler
      device handler delivers DATA to DEST_MAC
    }
}
END
```

10.3.3 SDN controller

We use a Floodlight controller to receive and send packets to switches. On receiving a packet, the controller extracts information such as SRC_MAC address, DEST_MAC address, ingress port, type of packet, SRC_IP address, and DEST_IP address. Based on the type of message it receives from the home network, the controller sends appropriate instructions to the switches and Gateway HomeBoxes, as shown in Fig. 10.9. It is also responsible for forwarding packets to switches and decides actions to be taken for a packet. It does so using a topology manager that discovers the set of switches to which the packets are to be forwarded. It can add, delete, and modify entries to the switch by sending an OpenFlow message.

Scheduler: queues the requests and schedules them

Packet Analyzer: analyzes the packet, retrieves device details from the database, and changes the packet format based on the type of service

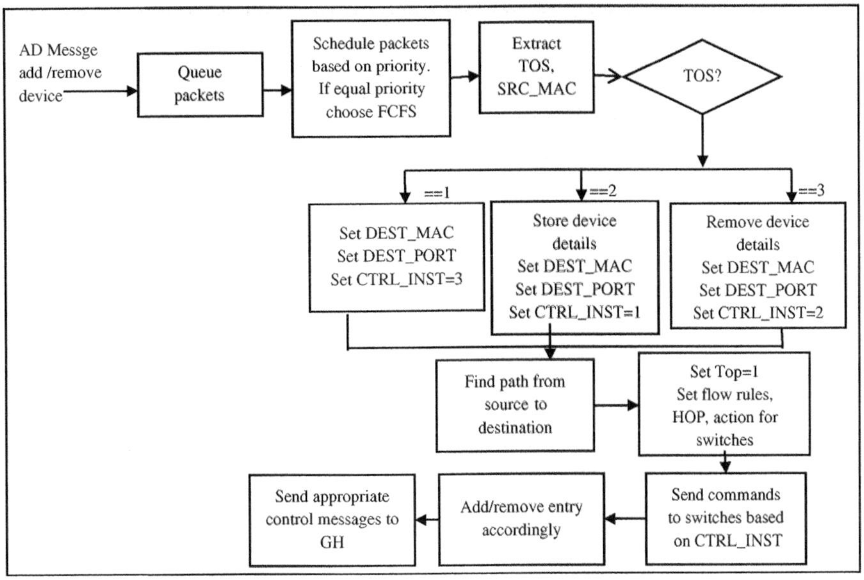

FIG. 10.9

Block diagram of controller.

Topology Manager: network view of the switches and hosts; selects the paths that connect source and destination

Forwarding: forwards the packet for flow setup to the switches along the paths

Optimized Flow Replacement Manager: buffers the recent flow entries of the switches and selects the flow rule to be removed based on the LRU replacement policy

INPUT:

```
AD PACKET, ADDev PACKET, REMDev PACKET,
```

OUTPUT: `Send Control Instructions to Switches and Gateway HomeBoxes`

ALGORITHM:

```
queue the packets in QUEUE[],
if(PACKET1.priority>PACKET2.priority) then
      send PACKET1 to packet processer
   else if(PACKET1. Priority=PACKET2.Priority)then
if(PACKET1.Duration>PACKET2.Duration) then
            send PACKET2 to packet processor
Extract TOS, SRC_MAC
If(TOS==1)
{   Search DEV_ID details in database
    set DEST_MAC in MAT_RULE
    set METADATA=CONFIG_SRCDEV_DESTDEV
    set DEST_PORT
    set CTRL_INST=03
```

Continued

```
}Else if(TOS==2)
{    STORE DEV_ID in database
     STORE DEV_DET in database
     STORE DEV_TOP in database
     Based on the DEV_OP performed by devices
     find all possible combinations of configurations that can be
       done
     set DEST_MAC in MAT_RULE
     set METADATA=ADD_DEV
     set DEST_PORT
     set CTRL_INST=01
}
Else if(TOS==3)
{    Remove DEV_ID details from database
     set DEST_MAC in MAT_RULE
     set METADATA=REM_DEV
     set DEST_PORT
     set CTRL_INST=02
}
Send SRC_IP & DEST_IP to topology manager
find the shortest Path between SRC_IP and DEST_IP
If(TOS==3)
{   for(switches.path)
   {forwardREMDev packet
     Remove FLOWTABLE.entry
     Set TOP=1;
   }
}Else if(TOS==2||TOS==1))
{ for(switches.path)
  {set HOP
   set ACTION
   Set TOP=1;
  }
}
Send CONTROL MESSAGES based on CTRL_INST value
END
```

10.3.3.1 Optimal rule replacement manager

We create a module called the Optimal Rule Replacement Manger that is responsible for handling replacements in the switch. Replacement takes place even before the flow table becomes full. The LRU replacement policy is used. Whenever the controller receives a new packet, it stores the packet information in a buffer. The packet information stored in the buffer includes match rule of a flow, priority, duration, packet count, and cookies. The block diagram for this module is as shown in Fig. 10.10. When the Optimal Rule Replacement Manager is invoked, the flow with a maximum duration and minimum packet count is chosen for replacement from the buffer.

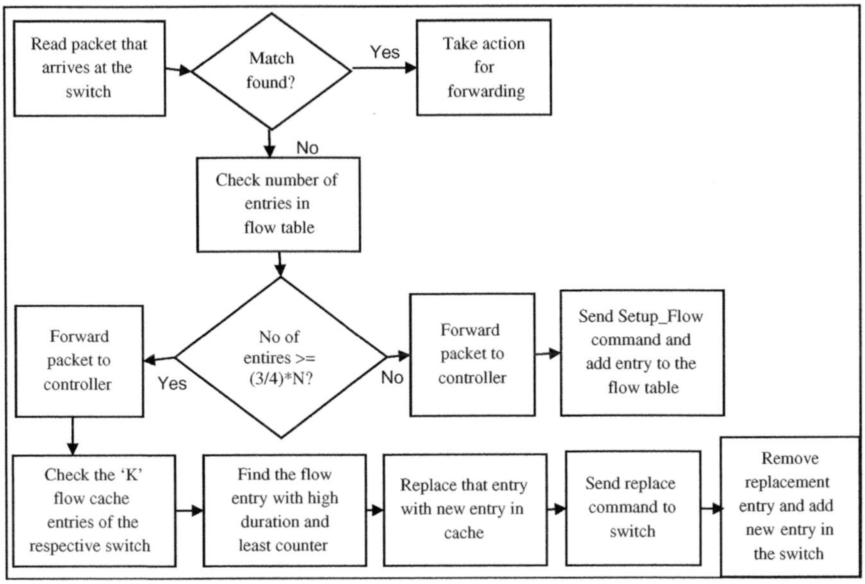

FIG. 10.10

Block diagram of optimal rule replacement manager.

INPUT:
Packets to switch, AD packet, ADDevpacket, REMDev packet, DS packet, DT packet
OUTPUT: ADD entry to flowtable
Algorithm: OptimizedFlowRep()

```
Extract the MATCH_RULE from the packets
If(FlowTable.entry==match)
    Call FORWARDpackets()
Else
{   check P, number of entries in the Flowtable
    if(p>=(3/4)*N)          //Total no of entries in the Flowtable
      { Forward Packet to Controller
        Extract MATCH_RULE,
        for(K Flow-Cache-Entries of Respective-switches)
        { if(DURATION=MAX && COUNTER=MIN)
            {   Choose that Entry for REPLACEMENT
                REPLACEMENT that Entry in the cache with new.Entry
            }
        } Send REPLACE command to Switch
        Replace FLOWTABLE.Entry with new.Entry
    }Else
    {       Forward Packet to Controller
            Send SETUP_FLOW Command to switch and add entry to switch
    }
}
End
```

10.3.4 Security module

The security module implemented at the controller side is the IDS that is responsible for detecting malicious attacks. Intrusion detection consists of two main processes, namely, the learning phase and real-time simulation. In the learning phase, the KDD CUP99 dataset is used for training the model. Using PACKET_IN message at the controller side, only features such as SRC_IP, DEST_IP, IP header length, flags, TCP flags, payload, total length, TOP, TOS, and version are extractable. Therefore only certain features from KDD CUP99 are used for training the IDS. The models are generated using three algorithms: (1) Bagging Ensemble of ID3, (2) Bagging Ensemble of Fuzzy Logic, and (3) DNN.

A decision tree is generated in the case of Bagging Ensemble of ID3, a fuzzy rule set is generated in the case of Bagging Ensemble of Fuzzy Logic, and a neural network is generated in the case of DNN. In real-time simulation, the packets sent by the hosts in the home network are sent to the switch. If any match is found in the flow table, then appropriate action is taken. If no match is found, then it is forwarded to the controller. When a PACKET_IN message is encountered by the controller, the packet is sent to the security module. The details are shown in Fig. 10.11.

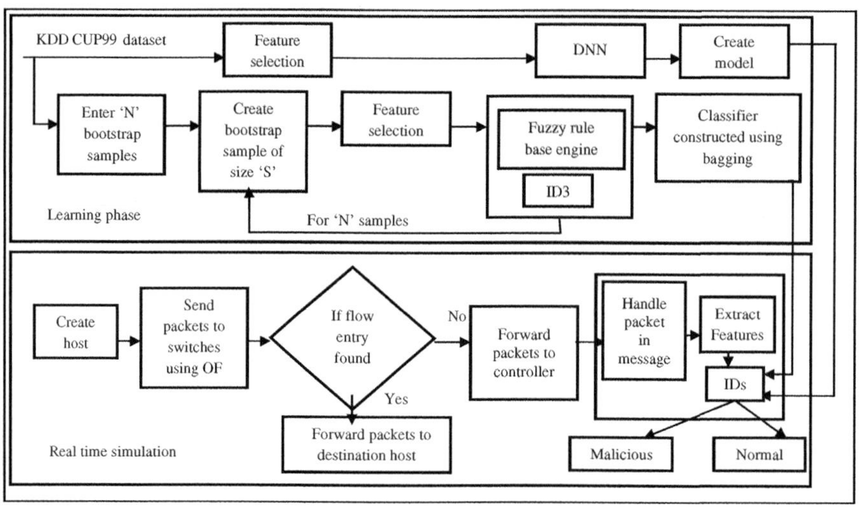

FIG. 10.11

Intrusion detection system using bagging ensemble technique and deep learning.

The security module extracts the necessary features from the PACKET_IN message and sends it to each of the algorithms. Each algorithm is run sequentially. The predictions given by each algorithm for the same packet are shown in Fig. 10.16. If the packet is malicious, then the attack is reported to the controller. The entire process is diagrammatically represented in Fig. 10.11. The controller takes necessary actions by disconnecting the host from the network that sent the attack.

10.3.4.1 Header extraction

Features related to the SDN environment are extracted from the PACKET_IN message. Features extracted are provided as inputs to Bagging of ID3, Bagging of Fuzzy Logic, and DNN submodules.

The switch forwards the packet to the controller if a match is not found in the flow table. The header information such as SRC_IP, DEST_IP, IP header length, total length, TCP flags, urgent pointer, TOP, TOS, and differential service field are extracted. IP header length and total length are used to compute the source bytes. Fig. 10.12 diagrammatically represents the header extraction process.

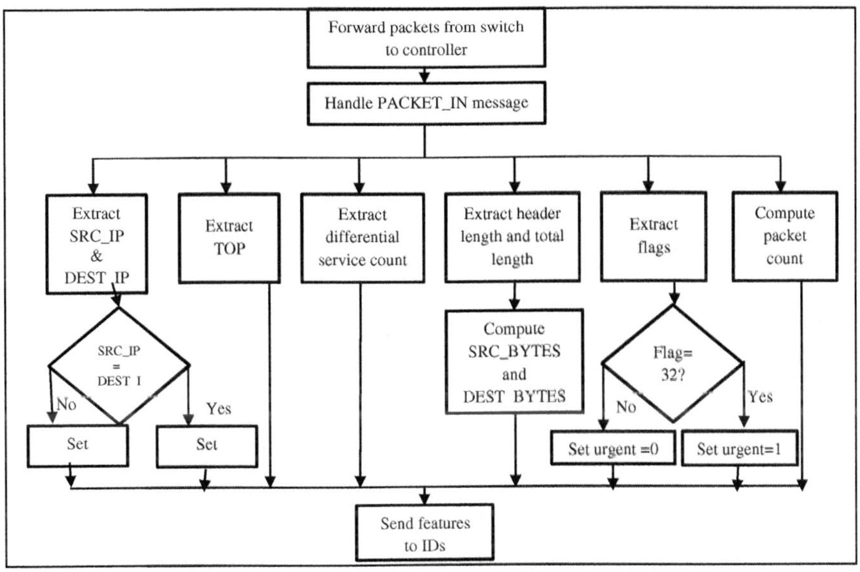

FIG. 10.12

Header extraction process at the controller side.

INPUT: `PACKET_IN message`
OUTPUT: `TOP, Src_bytes, dest_bytes, urgent, PKT_COUNT, Diffserv, land, flags`
ALGORITHM:

```
Extract TOP
Extract SRC_IP and DEST_IP
    If (SRC_IP=DEST_IP)
        set Land=1
    Else set Land=0
Extract flags
    If (TOP=TCP and TCP_flags=32 or TCP_flags=20)
        set Urgent=1
    Else set Urgent=0
```

Continued

```
    If (TOP=TCP and TCP_flags=2)
        set flags="S0"
    Else if (TOP=TCP and TCP_flags=3)
        set flags="SF"
    Else if (TOP=TCP and TCP_flags=4 or TCP_flags=6)
        set flags="REJ"
    Else set flags=0
Extract IP_HDR_LEN and TOT_LEN
    If (TOP=TCP)
        set Src_bytes=TOT_LEN-20
      Else if (TOP=UDP or TOP=ICMP)
        set Src_bytes=TOT_LEN-8
    Src_bytes=Src_bytes-IP_HDR_LEN
    If (TOP=ICMP and ICMP_Type=echo_reply)
       Set dest_bytes=TOT_LEN-IP_HDR_LEN-8
    Else set dest_bytes=0
Extract diffserv and compute diffserv count
Compute PKT_COUNT
```

10.3.4.2 Bagging ensemble of fuzzy logic

Bootstrap samples of KDD CUP99 dataset of specified size are created. Fuzzy classifier is chosen as base classifier. The process of bagging remains the same for both Bagging Ensemble of ID3 and Bagging Ensemble of Fuzzy Logic as represented in Fig. 10.13. Three bootstrap samples are generated with which three models are trained. Each model is trained using Fuzzy Unordered Rule Induction Algorithm (FURIA). Process such as quantization, normalization, fuzzy mapping is performed to form the fuzzy set. Quantization is performed to change the nonnumeric values to numeric values. Bootstrap samples are classified separately by the classifiers and majority of the vote is chosen. This is diagrammatically represented in Fig. 10.13.

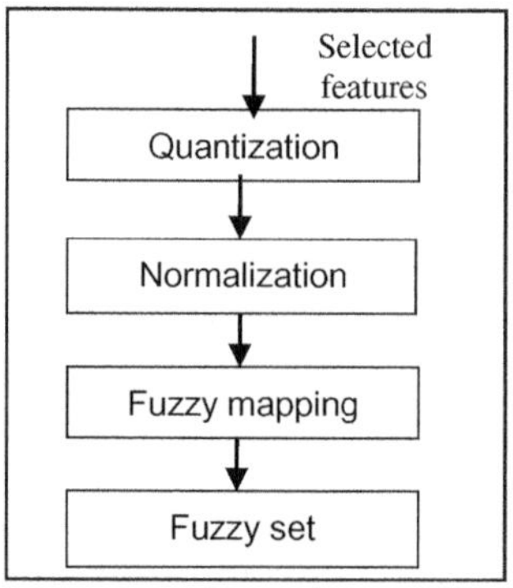

FIG. 10.13

Fuzzy rule base engine.

INPUT: Selected features from the KDDCUP99 dataset
OUTPUT: Creation of Fuzzy set
ALGORITHM:

```
Generate_rule()
        {    RECV (CHOSEN_ATTRIBUTES)
             Compute_Dev_Range (ATTR, NRML, ATCK)
             Gen_Defrule (ATTR, DEV_NRML, DEV_ATCK)
             Gen_Indefrule (ATTR, DEV_NRML, DEV_ATCK)
        }
Gen_Defrule (DEV_NRML, DEV_ATCK)
        { Assume {MAX, MIN} for NRML {1, 5}
              {MAX, MIN} for ATCK {2, 8}
                IF ATTR1 > 5, then data is "ATTACK"
                IF ATTR1 < 2, then data is "NORMAL".
        }
Gen_Indefrule (DEV_NRML, DEV_ATCK)
        {    Assume {MAX, MIN} for NRML {1, 5}
                {MAX, MIN} for ATCK {2, 8}
                  IF ATTR1<2 and ATTR1>5, then data is "NORMAL or ATTACK"
        }
```

Bagging_FuzzyLogic:

```
Bagging_FuzzyLogic (Examples, Attributes)
 {   Set M=3;
     Set BS=400;
     Select instances randomly for all the bootstrap samples
     SELECT_FEATURES (PKT)
     ASSIGN (FEATURE, FUZ_VAR)
     TRAPEZOIDAL (FUZZY_VAR)
       FLAG=Compare (INPUT_FUZ_SET, RULE_BASE)
       If (FLAG==L)
              REPORT (NRML, C)
       Else
              REPORT (ATCK, PATTERN, C)
 }
```

10.3.4.3 Bagging ensemble of ID3

Create bootstrap samples of KDDCUP99 dataset of specified size. ID3 classifier is chosen as base classifier ID3 algorithm classifies the packets as normal or malicious by using a decision tree. Bootstrap samples are classified separately by the classifiers and the majority of the vote is chosen. Three bootstrap samples are generated. Each bootstrap sample is used to build separate models. A decision tree is constructed by calculating the entropy and information gain for each attribute [30]. The attribute with high performance information gain is selected as the decision node. Hence instances are split based on the decision node. The process of calculating the information gain and entropy continues for both the left and right subtrees. When the node reaches the class, then it is marked as a leaf node. The predictions of each of the models are aggregated and the majority of the vote is chosen. Fig. 10.14. diagrammatically represents how ID3 is used to create a model.

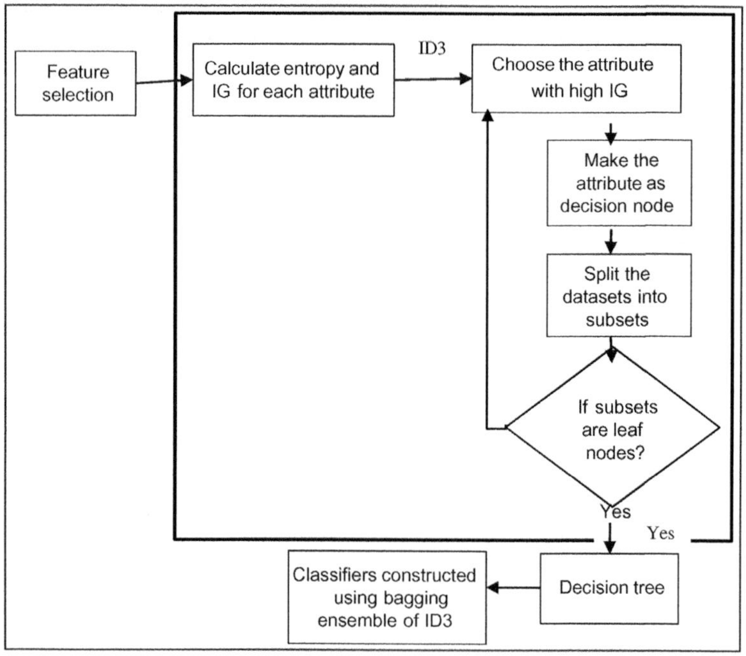

FIG. 10.14

Creation of model using ID3.

```
Input: Selected few Features from KDD CUP99 dataset
Output: Decision Tree (DTree)
Algorithm: ID3 (Examples, Attributes)
 Create ROOT.NODE for DTree
   Calculate ENTROPY of each Selected Attribute
    Calculate Information Gain for each attribute
            Gain(T,X)=Entropy(T)-Entropy(T,X)
    SPLIT into SUBSETS using Attribute with MIN_ENTROPY
    INSERT into DTree NODE containing that Attribute
   Recursive on SUBSETS using Pending Attributes.
```

10.3.4.4 Deep neural network

This module uses DNN to train datasets and generate a neural network model for intrusion detection. The input to DNN is a dataset in *.csv format. The data from the *.csv file is loaded into the matrix object. Once the neural network parameters are set, the attribute values are normalized so that it becomes easy to calculate. The input is then divided into batches that are assigned to separate threads.

Each batch takes 10 instances each and processes them. Cost and gradient value are computed each iteration and based on the gradient value, the instances are reduced in the next iteration. This process continues until the cost of the previous

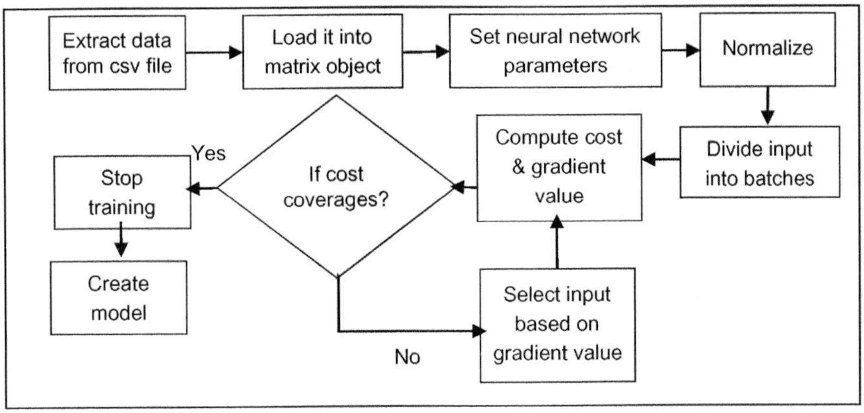

FIG. 10.15

Creation of model using DNN.

and current iterations converge. If the cost converges, training is stopped and the resultant model is used for detection process. Fig. 10.15 represents the creation of a model using DNN.

```
INPUT: Selected features from KDD CUP99 dataset
OUTPUT: A Neural Network (NN) Model
ALGORITHM: DNN (Examples, Target_Attribute, Attributes)
 Extract data from CSV File
 Matrix M= Load from SDN_DATASET.csv
    SET NNParams
     numHiddenLayers, PoolWidth, Pool Height
 Normalize(Attribute Values)
    Repeat for each Weight Update Iteration of NODES
      Compute COST and GRADIENT_VAL.
    until COST Converges
 Store the NN model.
```

10.3.5 Implementation

10.3.5.1 Tools used

The following paragraphs describe the tools used.

Mininet is a network emulator that creates a network of virtual hosts, switches, controllers, and links. Mininet hosts run standard Linux network software, and its switches support OpenFlow for highly flexible custom routing and SDN [31].

A Floodlight controller is an SDN controller offered by Big Switch Networks that works with the OpenFlow protocol to orchestrate traffic flows in a SDN environment. OpenFlow is the first and most widely used SDN standard; it defines the open communications protocol in a SDN environment that allows it to make changes to the network. The SDN controller is responsible for maintaining all of the network rules and providing the necessary instructions to the underlying infrastructure on how traffic should be handled [32].

Hping3 is a CLI TCP/IP packet assembler/analyzer. It supports TCP, UDP, ICMP, and RAW-IP protocols, has a trace route mode, the ability to send files between a covered channel, and many other features [33].

10.3.5.2 Connection establishment and network creation

The aforementioned tools are used to set up a home network in a SDN. In Fig. 10.2, the SDN controller and hosts are set up representing the device nodes in home networks. The following section describes the installation and creation of the network. The connection establishment among the hosts and the controller are detailed with Ubuntu scripts and the tools used.

1. ***Installation steps***

```
Step 1: Install Virtual Box
Step 2: Download Mininet VM from http://mininet.org/download/ and
    install the VM in virtual box.
Step 3: Download Floodlight VM from https://floodlight.atlassian.
    net/wiki/display/floodlightcontroller/Floodlight+V
    Step 3.1: Start it in Virtual Box
Step 3.2: Open Terminal in the floodlight VM and execute the
    following commands
    floodlight$ cd ~/floodlight
    floodlight$ git pull origin master
    floodlight$ git submodule init
    floodlight$ git submodule update
    floodlight$ ant
    floodlight$ sudomkdir /var/lib/floodlight
    floodlight$ sudochmod 777 /var/lib/floodlight
Step 3.3: once the commands in step 3.2 are executed, then build
    and run Floodlight within the VM by executing the following
    commands
    floodlight$ java -jar target/floodlight.jar
    floodlight$ ant eclipse
Step 3.4: Now open floodlight in eclipse
    Step 3.4.1Open eclipse and create a new workspace
    Step 3.4.2 File -> Import -> General -> Existing Projects
    into Workspace. Then click "Next".
    Step 3.4.3 From "Select root directory" click "Browse".
    Select the parent directory where you placed
    floodlight earlier.
    Step 3.4.4 Check the box for "Floodlight". No other Projects
    should be present and none should be
    selected.
    Step 3.4.5 Click Finish.
    Step 3.4.6 Run the floodlight from eclipse by clicking Run.
```

2. *Creation of network and connection establishment*

```
Step 1: In Mininet virtual machine, execute the following command
    Mininet$ sudo ~/mininet/examples/miniedit.py
     This command will open the MiniEdit GUI. We can create as many
     numbers of hosts, virtual links, switches and controllers using
     this GUI. This is depicted in Fig. A1.1.
Step 2: Create a controller. Edit controller conFigureurations and
    set the IP address of controller to that of the IP address of
    Floodlight VM and set the port number to 6653.
Step 3: Create switches and hosts. Set the name of switches. Set
    the name and IP address of hosts.
Step 4: Go to file->Preferences ->check the "Start CLI".
Step 5: Run floodlight in eclipse in the floodlight VM
Step 6: In Mininet VM click "Run" in MiniEdit. This will open
    Mininet CLI interface and displays that the network is
    created successfully.
Step 7: To check for connection establishment. Send packets from
    mininet CLI interface by executing the following command
    Mininet$ h1 ping h2
     (This is depicted in Fig. A1.3)
Step 7: In eclipse running in Floodlight VM, If you can
    successfully view the IP address of h1 and h2 in the console,
    then connection establishment is successful.
     (This is depicted in Fig. A1.2)
```

10.3.5.3 Dataset description

The renowned dataset KDD CUP99 is used in training the IDS. Though this dataset has limitations, it is a widely used benchmark dataset for the evaluation of anomaly detection methods. Moreover, datasets are not available in particular for SDN, which is why even recent works in Network Intrusion Detection Systems (NIDS) rely on KDD CUP99 and NSL_KDD [19, 20, 34].

The attacks in the dataset, fall under four main categories:

1. DoS (e.g., syn flood)
2. R2L unauthorized access from a remote machine (e.g., guessing password)
3. U2R unauthorized access to local superuser (root) privileges (e.g., various "buffer overflow" attacks)
4. Probing surveillance and other probing (e.g., port scanning)

Initially a precise dataset of 600 records containing SDN adapted features is collected. The dataset contains both normal network behaviour as well as two types of attacks: DoS and Probe. This is also relevant to IoT, as the home devices are connected in a network similar to commonly participating networks. The dataset also contains records that are collected from the SDN network for both normal and abnormal behavior. The dataset was created in *.arff format. The attributes in the dataset are described in Table 10.1. In particular the data in flag field and their description are given in Table 10.2. This is relevant to SDN for home network devices concerning security of networks.

Table 10.1 Attribute description in collected dataset

S. No	Feature	Description
1.	Land	Set to 1 if SRC_IP and DEST_IP are same else 0
2.	Protocol	type of the protocol, e.g., tcp, udp, etc.
3.	Flags	normal or error status of the connection
4.	Source bytes	number of data bytes from source to destination
5.	Destination bytes	number of data bytes from destination to source
6.	Urgent	number of urgent packets
7.	Count	number of connections to same host as the current connection withi past two seconds
8.	Diffserv	number of connections to different services

Table 10.2 Flags field description

Flag	Description
REJ	connection rejected—initial
SF	normal SYN/FIN completion
S0	initial SYN seen but no reply
S1	connection established (SYN's exchanged), nothing further seen
S2	connection established, initiator has closed their side
S3	connection established, responder has closed their side
S4	SYN acknowledgement seen, but no initial SYN seen

10.3.5.4 Performing attacks

The few attributes, taken from the KDD CUP99 dataset as placed in Table 10.1, are more relevant for extracting needed information from the controller with security impacts. These attributes help in detecting some of the common attacks that are prevalent in IDS. Apart from these, the SDN controller induced various attacks that were sent using the hping3 tool, including Smurf, Neptune, IPsweep, and land attack. From this same network, for standard dataset-related security attacks, in some cases, if we know the protocol type and payload size an attack can be manufactured from these very details [35]. Similarly, if the protocol is UDP and payload size is way too large, then it is an amplification attack later generated into a DoS attack. So with the parameters from the dataset, few attacks are framed and few are generated in a SDN-based home network setup.

A Smurf attack [36] is a DDoS attack with IP spoofing. The attackers use ICMP echo request packets that are directed to the IP broadcast address from a remote location to create a DoS attack [36]. The attacker sends ICMP "echo request" packets to the broadcast address (xxx.xxx.xxx.255) of many subnets with the source address spoofed to be that of the intended victim. Any machines that are listening on these subnets will respond by sending ICMP "echo reply" packets to the victim.

The command to send a Smurf attack is as follows: *hping3 -1 –d 1032 –flood –a Victim_IP Broadcast_Addr* "−1" indicates that it is an ICMP packet and "d" is used to specify the data size. Victim_IP specifies the victim IP and "flood" is used to flood the packets. Broadcast_Addr specifies the broadcast address so that the attack is distributed.

A Neptune/SYN flood attack [36,37] is a DoS attack where every TCP/IP implementation is vulnerable. Here, the half-open TCP connection causes the tcpd server to add a record to the data structure that stores information describing all pending connections. This data structure is of finite size, and it can be made to overflow by intentionally creating too many partially open connections. The half-open connections data structure on the victim's server system will eventually fill and the system will be unable to accept any new incoming connections until the table is emptied out.

The command to send a Neptune attack is as follows: *Hping3 -S –flood -V Victim_IP* "S" sets the SYN field in TCP flags and "flood" is used to flood the packets. Victim_IP specifies the victim IP.

An IPsweep attack [38] is a probe attack that is performed to gather information about active hosts in a network. The attacker sends multiple ICMP echo request packets to the victim's IP. The attacker comes to know that the victim host is up if the host responds with an ICMP echo reply packet. This information about the host is gathered to send a future attack. The attack takes place soon without the victim knowing. The differential service is often set to immediate transfer so that the attack goes unnoticed.

The command to send an IPsweep attack is as follows: *hping3 -1 –d 0 –flood -o 1 Victim_IP* "−1" indicates that it is an ICMP packet and "d" is used to specify the data size. Victim_IP specifies the victim IP and "flood" is used to flood the packets. "o" specifies the differential service field. A value of 1 in the differential service field indicates the packet should be immediately transferred.

A land attack [39,40] is also a DoS attack of some older TCP/IP implementations. Here, it sends a single SYN packet with the source address spoofed to be the same as the destination address. When a TCP SYN packet with an identical source and destination address is received by the host, the system completely locks up. In order to restore service, the machine has to be physically turned off and on again.

The command to send a land attack is as follows: *hping3 –V –d 120 –S –flood –a Spoofed_IPVictim_IP* "d" is used to specify the data size. Victim_IP specifies the victim IP and "flood" is used to flood the packets. "a" specifies that the IP is spoofed. Here the spoofed_IP and Victim_IP will be the same.

10.4 Results

10.4.1 Evaluation parameters

A simple star topology network was created that consisted of four hosts, namely, h1, h2, h3, and h4 in home network 1 and another four hosts, namely, h5, h6, h7, and h8 in home network 2. The default data rate is 64 bytes/sec for each packet sent. The data rate can be changed manually in the hping3 tool. Fifty packets were sent in real time and their packet features were tracked by storing the feature values in a file. Out of the 50 packets, 40 were malicious and 10 were normal. Attacks were sent using the hping3 tool. Table 10.3 lists the evaluation parameters. These give the fragmentation, flow management, and security evaluation parameters for SDN home networks.

Table 10.3 Evaluation parameters

Parameter	Value
No. of hosts	8
No. of Switches	2
No. of Controller	1
Topology	Star
Data Rate	64 bytes/sec
No. of Packets Sent	50
Type of Packets sent	ICMP, TCP and UDP
No. of TCP packets	22
No. of UDP packets	3
No. of ICMP packets	25

10.4.2 Evaluation metrics

(i) Metrics for data sharing [41]:

Bandwidth estimation and rate adjustment for video data are given in Eq. (10.2)

$$b_i = \sum d_i / t_4 - t_1 \tag{10.2}$$

Exponentially averaging the samples using the formula given in Eq. (10.3),

$$B_i = \alpha^* b_{i-1} + (1-\alpha)(b_i + b_{i-1})/2 \tag{10.3}$$

where b_i, achieved sample rate (amount of data/inter arrival time between packets)

d_i, data packet with d1, d2, d3 and d4 bytes
α, weighting parameter
t, timeperiod
α close to 1 implies best performance

Roundtrip is given in Eq. (10.4) [41]:

$$RTT = (\alpha \cdot Old\_RTT) + ((1 - \alpha) \cdot New\_Round\_Trip\_Sample) \quad (10.4)$$

where α is constant weighting factor ($0 \leq \alpha < 1$). RTT should be minimum for better performance.

(ii) Metrics for controller performance:

Flow setup latency: Time taken to set up flow in a flow table.

This depends on the quality of the controller.

(iii) Metrics for flow modification rate are given in Eq. (10.5)

Space Complexity: If Nis the number of switches managed by a controller in a current network, K is number of entries in the flow buffer, B is the bytes occupied by each flow entry then,

$$\text{Space complexity} = N*B*K \text{ bytes} \quad (10.5)$$

K should be chosen in such a way that the storage is optimal

Flow modification ratio is given in Eq. (10.6): if k1 is the number of modifications done for P packets and k2 is the number of modifications performed by the proposed algorithm for the same P packets, then modification ratio is given by

$$\mu = M(k1, P)/M(k2, P) \quad (10.6)$$

if $\mu > 1$, the algorithm performs better

(iv) Accuracy is shown in Eq. (10.7)

Accuracy is the proportion of the total number of predictions that were correct. The higher the accuracy, the better the algorithm performs.

$$\text{Accuracy} = (TP + TN)/(TP + TN + FP + FN) \quad (10.7)$$

where TP is True Positive, which is the number of actual malicious packets that were correctly classified as malicious;

FP is False Positive, which is the number of actual normal packets that were incorrectly classified as malicious;
FN is False Negative, which is the number of actual malicious packets that were incorrectly classified as not malicious; and
TN is True Negative.

(v) Precision is shown in Eq. (10.8):

Precision gives the proportion of the predicted positive cases that were correct.

$$\text{Precision} = TP/(TP + FP) \quad (10.8)$$

(vi) Recall is shown in Eq. (10.9):

Recall or sensitivity is the proportion of the positive cases that were correctly identified.

$$\text{Recall} = \text{TP}/(\text{TP} + \text{FN}) \quad (10.9)$$

(vii) F-measure is shown in Eq. (10.10):

The F-measure is the harmonic mean of recall and precision.

$$\text{F-measure} = 2{*}\text{Precision}{*}\text{Recall}/(\text{Precision} + \text{Recall}) \quad (10.10)$$

(viii) Attack Accuracy is shown in Eq. (10.11):

Accuracy checks whether it correctly identifies the different attack types individually.

$$\text{AttAcc} = \frac{1}{C-1} {*} \sum\nolimits_{i=2}^{C} \text{Recall}_i \quad (10.11)$$

where C is the number of classes. The first class (i = 1), is considered to be the normal class.

(ix) Attack Detection Rate is shown in Eq. (10.12):

Attack detection rate is the accuracy rate for the attack classes.

$$\text{Attack Detection Rate} = \frac{\sum\nolimits_{i=2}^{c} \text{TP}(i)}{\sum\nolimits_{i=2}^{c} \text{TP}(i) + \text{FN}(i)} \quad (10.12)$$

(x) False Alarm Rate is shown in Eq. (10.13):

The false alarm rate checks the percentage of "false negatives" found, that is, those instances identified as "alarms" but that are actually normal behavior.

$$\text{FAR} = \text{FP}/(\text{FP} + \text{TN}) \quad (10.13)$$

(xi) Root mean square error

The root-mean-square deviation (RMSD) or root-mean-square error (RMSE) is used to measure the differences between values (sample and population values) predicted by a model or an estimator and the values actually observed.

$$\text{RMSE} = \sqrt{\frac{\sum\nolimits_{t=1}^{n} (y'(i) - y(i))\hat{2}}{n}} \quad (10.14)$$

where y'(i), Predicted value for ith packet

y(i), observed value for ith packet
n, total number of packets

These are the listed evaluation parameters used to measure the performance of the created SDN-based home network.

10.4.3 Experimentation results

10.4.3.1 Test cases

Table 10.4 lists the packet behaviors of packets that were selected for evaluation. It displays the count of each type of packets that belong to each of the classes, including normal and attack types. Packets were sent from hosts in the Mininet side to the controller. Table 10.5 shows the set of test cases that were passed and their actual outputs recorded. Each module was tested separately. Table 10.6 displays the attack types done on different time frames. The time frame describes the starting time of the attack that was simulated to the time the attack was detected. The variations of different time frames for each attack type are due to the amount of time taken for processing the packet features. For example, to detect a land attack, finding the SRC_IP and DEST_IP were enough. To detect a Smurf attack, TOP and Src_bytes were enough.

Table 10.4 Packet behaviors of packets sent for evaluation

	Class				
	Normal	Smurf	Neptune	IPsweep	Land
Number of TCP packets	3	0	11	0	9
Number of UDP Packets	2	0	0	0	0
Number of ICMP packets	5	11	0	9	0

Packet features such as land, TOP, flags, source bytes, destination bytes, differential service, urgent field and count for each of the 50 packets are recorded in Table 10.7. The predicted values by each of the algorithms (i.e., Bagging Ensemble of ID3, Bagging Ensemble of Fuzzy Logic, and DNN) are shown Fig. 10.16. TP, FP, TN, and FN values are calculated for each of the algorithms and recorded and shown in Fig. 10.17.

Accuracy, precision, and recall values are calculated using Eqs. (10.7)–(10.9) based on the recorded values of TP, FP, TN, and FN for each of the algorithms. The precision and recall values are as shown in Figs. 10.18 and 10.19, where precision is the ratio of correct, positive classifications to the number of actual positives, and recall is the ratio of the number of correct positive classifications out of those that were classified as positive (i.e., sensitivity). The F-Measure is calculated based on Eq. (10.10) based on recorded values of precision and recall, which are compared in Fig. 10.20. Attack detection rate, F-Measure, and precision-recall curve are plotted in respective graphs. Attack accuracy and overall accuracy are represented in a bar graph in Fig. 10.21.

Table 10.5 Test cases for IDS

TEST ID	Module	Objectives	Input	Expected output	Actual output
ID_1	Controller	Check whether modules are loaded properly	Launch controller	Display “initialization successful” for respective modules	Initialization successful
ID_2	Controller	Check whether packets are delivered to the destination host	Send packet from one host to another	Display “Packet Received successfully” at controller side else display “Destination host unreachable”	Packet received successfully
ID_3	Bagging_ID3	Number of Bootstrap samples; N should be positive	$N < 0$	Display “Bootstrap sample value should not be negative, Enter a positive value”	Bootstrap sample value should not be negative
ID_4	Bagging_ID3	Bootstrap sample size; BS should not exceed dataset size S	$BS > S$	Display “ Bootstrap sample size should not be greater than dataset size”	Bootstrap sample size should not be greater than dataset size
ID_5	Bagging_ID3	Bootstrap sample size; BS should be positive	$BS < 0$	Display “Bootstrap sample size should be positive”	Bootstrap sample size should be positive
ID_6	Bagging_FuzzyLogic	Number of Bootstrap samples; N should be positive	$N < 0$	Display “Bootstrap sample value should not be negative, Enter a positive value”	Bootstrap sample value should not be negative, Enter a positive value
ID_7	Bagging_FuzzyLogic	Bootstrap sample size; BS should not exceed dataset size S	$BS > S$	Display “ Bootstrap sample size should not be greater than dataset size”	Bootstrap sample size should not be greater than dataset size

ID_8	Bagging_FuzzyLogic	Bootstrap sample size; BS should be positive	S<0	Display "Bootstrap sample size should be positive"	Bootstrap sample size should be positive
ID_9	Bagging_ID3	Dataset file must be available	No file	Display "File not found"	File not found exception
ID_10	Bagging_ID3	Only .arff file should be used for training	Any file other than .arff file	Display "Dataset should be a .arff file"	Dataset should be a .arff file
ID_11	DNN	Only .CSV file should be used for training	Any file other than .CSV file	Display "Dataset should be a .CSV file"	Dataset should be a .CSV file
ID_12	DNN	Learning rate, L, should be positive	L<0	Display "Learning rate can't be negative"	Learning rate can't be negative
ID_13	DNN	Batch_size, BS, should be 1 for convolution layers	BS<0	Display "Batch size cannot be negative"	
ID_14	DNN	Batch_size, Ba_S, should be 1 for convolution layers	Any positive value other than 1	Display "Batch size should be 1 for convolution layers"	Batch size should be 1 for convolution layers

Table 10.6 Time frame of each attack type

Attack type	Starting time (in ms)	Ending time (in ms)
Smurf	0	20
Neptune	0	25
IPsweep	0	23
Land	0	18

Table 10.7 Test set containing packet feature details and class label

Packet No.	Land	Protocol	Flag	Src bytes	Dest bytes	Urgent	Diff Serv	Count	Class
1	0	Udp	0	315	0	0	16	1	Normal
2	0	Icmp	0	64	0	0	0	1	Normal
3	0	Tcp	SF	720	0	0	9	1	Normal
4	0	Tcp	S0	120	0	0	9	1	Normal
5	0	Tcp	S0	121	0	0	10	1	Normal
6	0	Icmp	0	60	0	0	0	1	Normal
7	0	Icmp	0	1040	0	0	0	1	Smurf
8	0	Icmp	0	1032	0	0	0	1	Smurf
9	0	Icmp	0	1031	0	0	6	1	Smurf
10	0	Icmp	0	1033	0	0	7	2	Smurf
11	0	Icmp	0	1040	0	0	6	3	Smurf
12	0	Icmp	0	1040	0	0	6	4	Smurf
13	0	Icmp	0	1040	0	0	6	5	Smurf
14	0	Tcp	S0	15	0	0	6	5	Neptune
15	0	Tcp	S0	12	0	0	7	8	Neptune
16	0	Tcp	S0	120	0	0	7	9	Neptune
17	0	Tcp	S0	120	0	0	0	10	Neptune
18	0	Tcp	S0	68	0	0	0	6	Neptune
19	0	Tcp	S0	0	0	0	9	8	Neptune
20	0	Tcp	S0	12	0	0	9	9	Neptune
21	0	Icmp	0	8	0	0	1	5	Ipsweep
22	0	Icmp	0	8	0	0	1	6	Ipsweep
23	0	Icmp	0	8	0	0	1	8	Ipsweep
24	0	Icmp	0	18	0	0	1	7	Ipsweep
25	0	Icmp	0	8	0	0	1	11	Ipsweep
26	1	Tcp	S0	0	0	0	16	1	Land
27	1	Tcp	S0	8	0	0	16	1	Land
28	1	Tcp	S0	12	0	0	15	2	Land
29	1	Tcp	S0	16	0	0	13	3	Land
30	1	Tcp	S0	16	0	0	12	1	Land

Table 10.7 Test set containing packet feature details and class label *Continued*

Packet No.	Land	Protocol	Flag	Src bytes	Dest bytes	Urgent	Diff Serv	Count	Class
31	0	Udp	0	60	0	0	16	1	Normal
32	0	Icmp	0	48	0	0	0	1	Normal
33	0	Icmp	0	1030	0	0	8	6	Smurf
34	0	Icmp	0	1030	0	0	8	7	Smurf
35	0	Tcp	S0	10	0	0	11	10	Neptune
36	0	Tcp	S0	16	0	0	6	4	Neptune
37	0	Icmp	0	12	0	0	1	12	IPsweep
38	0	Icmp	0	12	0	0	1	11	IPsweep
39	1	Tcp	S0	16	0	0	3	1	Land
40	1	Tcp	S0	64	0	0	16	1	Land
41	0	Icmp	0	56	0	0	0	1	Normal
42	0	Icmp	0	55	0	0	1	1	Normal
43	0	Icmp	0	1030	0	0	8	8	Smurf
44	0	Icmp	0	1033	0	0	8	9	Smurf
45	0	Tcp	S0	0	0	0	14	13	Neptune
46	0	Tcp	S0	15	0	0	7	4	Neptune
47	0	Icmp	0	14	0	0	1	13	IPsweep
48	0	Icmp	0	15	0	0	1	14	IPsweep
49	1	Tcp	SF	8	0	0	0	2	Land
50	1	Tcp	SF	64	0	0	16	2	Land

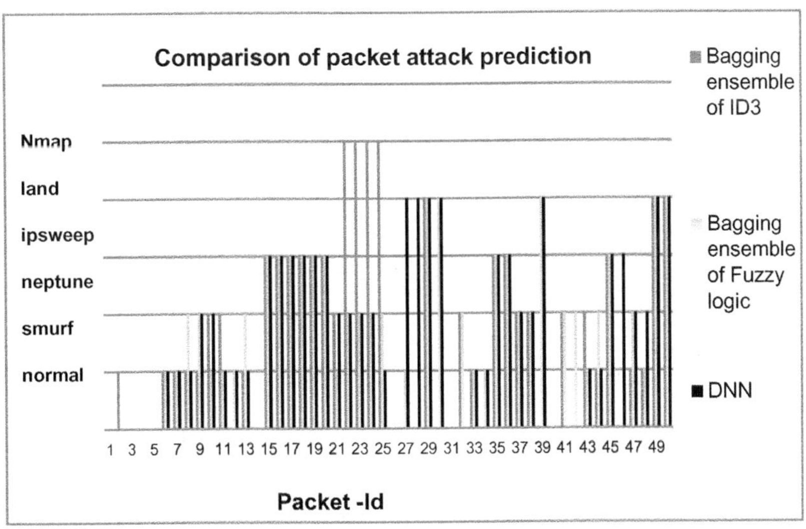

FIG. 10.16

Comparison of attack prediction by three algorithms.

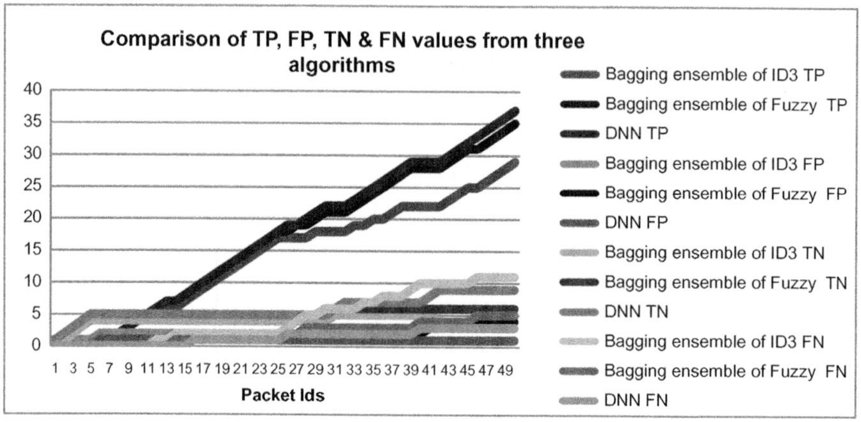

FIG. 10.17

TP, FP, TN, and FN values for three algorithms.

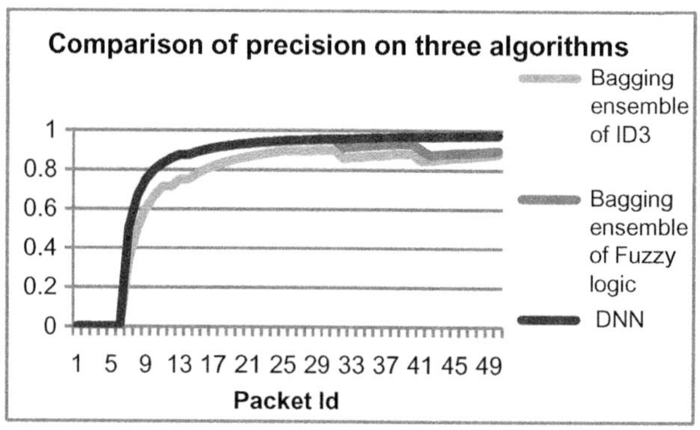

FIG. 10.18

Comparison of precision.

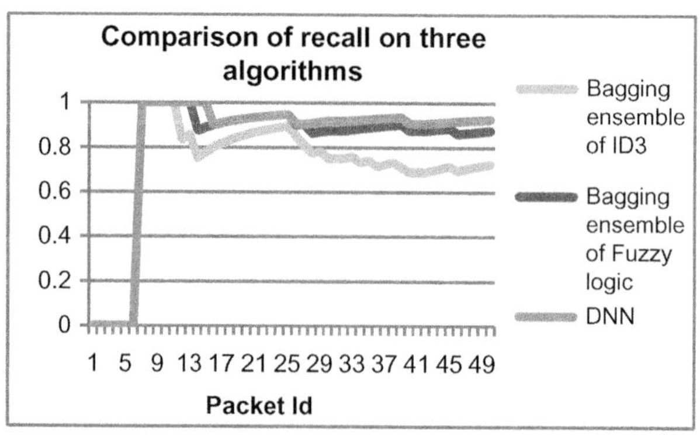

FIG. 10.19

Comparison of recall.

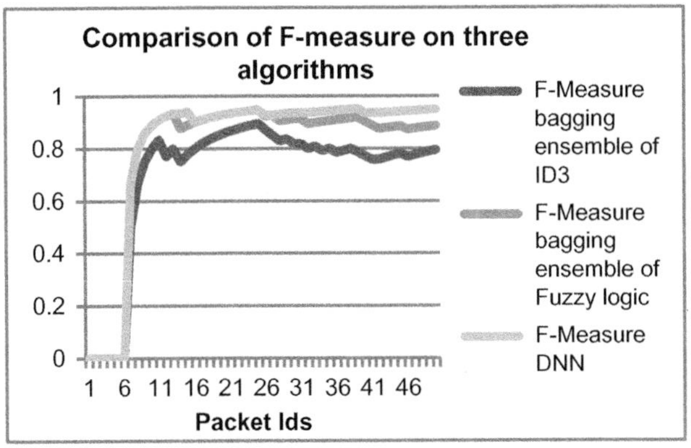

FIG. 10.20

Comparison of F-measure.

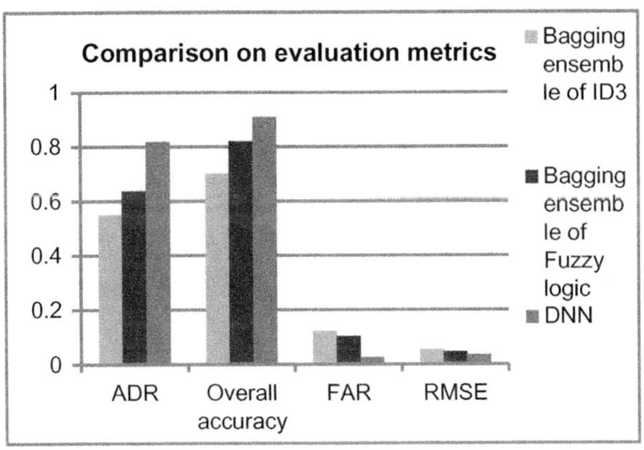

FIG. 10.21

Comparison of accuracy.

When sending a packet (normal/malicious), only the selected feature details of the packet are extracted and sent to the IDS module. The class label is not known. The IDS module predicts whether the packet is malicious or not. Results were taken by running each of the algorithms separately for the same packet and are shown in Fig. 10.16.

The attack accuracy for each algorithm is recorded and shown in Fig. 10.21. Attack Detection Rate (ADR), overall accuracy, false alarm rate (FAR), and RMSE are recorded in Fig. 10.21. From this table, it can be inferred that DNN has a low FAR of 0.026 compared to Bagging Ensemble of ID3 with a FAR of 0.121 and Bagging Ensemble of Fuzzy Logic with a FAR of 0.121. ADR and overall accuracy for DNN is higher than that of the other two algorithms. This indicates that DNN has a higher learning curve than the others. Similarly, DNN has a low RMSE of 0.034 compared to Bagging Ensemble of ID3 with a RMSE of 0.054 and Bagging Ensemble of Fuzzy Logic with a RMSE of 0.0447.

10.4.3.2 Bagging ensemble of ID3

The ADR is very low for the first six packets since they are normal packets. The ADR reaches its maximum value for the first attack it detects since the seventh packet is a Smurf packet. The ADR is high for the next three packets since it detected the malicious packets correctly. The second peak is encountered for 25th packet where the ADR for Bagging Ensemble of ID3 has a value of 0.895. For the subsequent packets, the ADR decreases and increases depending on whether the algorithm correctly detects the attack or not. On analyzing the last packet in the test set, the ADR reaches a value of 0.725. This is as shown in Fig. 10.21.

The F-Measure is very low for the first six packets since TP and FP values are zero, since it computes the mean of false positives and false negatives. Hence the precision and recall values were zero for the first six packets. The peak is obtained for 25th packet where the precision and recall values are 0.894 and 0.894, respectively. This is depicted in Figs. 10.18 and 10.19. The ADR for the 25th packet was also high at this point. For the last packet in the test set, the F-Measure reaches a value of 0.894.

The precision recall curve for Bagging Ensemble of ID3 is depicted in Figs. 10.18 and 10.19. Initially the precision and recall values are zero since the first six packets are normal packets. When the first attack was detected, the TP value was 1, the FP value was 2, and the FN value was 0. Hence the precision having a value of 0.33 was less compared to recall value of 1. This explains the initial increase of recall value. As the attack increases, the precision value of Bagging Ensemble of ID3 increases and its recall decreases. After analyzing the last packet, the precision reaches a value of 0.878 and recall reaches a value of 0.725.

10.4.3.3 Bagging ensemble of fuzzy logic

The ADR is very low for the first six packets since they are normal packets. The attack detection rate reaches its maximum value for the first attack it detects since the seventh packet is a Smurf packet. The ADR is high for the next seven packets since the true positive increases. There is a sudden decrease of ADR at the 14th packet since the false positive rate increases. The second peak is encountered for 25th packet where the ADR for Bagging Ensemble of Fuzzy Logic has a value of 0.947. For the subsequent packets, the ADR decreases and increases depending on whether the algorithm correctly detects the attack or not. On analyzing the last packet in the test set, the ADR reaches a value of 0.875. This is shown in Fig. 10.21.

The F-Measure is very low for the first six packets since TP and FP values are zero. Hence the precision and recall values were zero for the first six packets. The peak is obtained for 25th packet where the precision and recall values are 0.947 and 0.947, respectively. This is depicted in Fig. 10.20. The ADR for the 25th packet was also high at this point. For the last packet in the test set, the F-Measure stabilizes to a value of 0.885.

The precision recall curve for Bagging Ensemble of Fuzzy Logic is as shown in Figs. 10.18 and 10.19. Initially the precision and recall values are zero since the first six packets are normal packets. When the first attack was detected, the TP value was 1, the FP value was 1, and the FN value was 0. Hence the precision having a value of 0.5 was less compared to recall value of 1. This explains the initial increase of recall value. As the attack increases, the precision value of Bagging Ensemble of Fuzzy Logic increases and recall of Bagging Ensemble of Fuzzy Logic decreases. After analyzing the last packet, the precision reaches a value of 0.897 and recall reaches a value of 0.875.

10.4.3.4 Deep neural network

The ADR is very low for the first six packets since they are normal packets. The ADR reaches its maximum value for the first attack it detects since the seventh packet is a Smurf packet. The ADR is high for the next nine packets since the true positive increases. There is a sudden decrease of ADR at the 16th packet since the false positive rate increases. The second peak is encountered for the 25th packet where the ADR for DNN has a value of 0.947. For the subsequent packets, the ADR decreases and increases depending on whether the algorithm correctly detects the attack or not. On analyzing the last packet in the test set, the ADR reaches a value of 0.935. From the graphs in Figs. 10.16–10.21, DNN has a higher ADR of 0.935 compared to Bagging Ensemble of ID3 with an ADR of 0.725 and Bagging Ensemble of Fuzzy Logic with an ADR of 0.875.

The F-Measure is very low for the first six packets since TP and FP values are zero. Hence the precision and recall values were zero for the first six packets. The first peak is obtained for 15th packet where the precision and recall values are 0.888 and 0.888, respectively. The F-Measure for DNN slowly increases. This is depicted in Fig. 10.20. For the last packet in the test set, the F-Measure reaches a value of 0.949. From Figs. 10.16–10.21 it can be inferred that the harmonic mean between precision and recall is higher for DNN compared to Bagging Ensemble of ID3 with a F Measure of 0.794 and Bagging Ensemble of Fuzzy Logic with a F-Measure of 0.885.

The precision recall curve for DNN is depicted in Figs. 10.18 and 10.19. Initially the precision and recall values are zero since the first six packets are normal packets. When the first attack was detected, the TP value was 1, the FP value was 1, and the FN value was 0. Hence the precision having a value of 0.5 was less compared to recall value of 1. This explains the initial increase of recall value. As the attack increases, the precision value of DNN increases and recall of DNN decreases. After analyzing the last packet, the precision reaches a value of 0.973 and recall reaches a value of 0.925. It can be inferred that the precision values for all the three algorithms are greater compared to their respective recall values, though initially recall values were higher than their respective precision values. This means that the false positive rate for all the three algorithms is less compared to the false negative rate. This is what is actually required for an efficient IDS. Predicting a malicious packet to be normal is highly dangerous compared to predicting a normal packet to be malicious.

Among the 50 packets in the test set, 11 packets were Smurf packets. Bagging Ensemble of ID3 and Bagging Ensemble of Fuzzy Logic detected six and seven packets as Smurf among the 11 Smurf packets, whereas DNN detected nine packets as Smurf. Hence the attack accuracy for DNN is higher than the attack accuracy for Bagging Ensemble of ID3 and Bagging Ensemble of Fuzzy Logic. None of the Smurf packets were detected as normal by any of the three algorithms. The false negative rate for Smurf attacks is 0. Some Smurf attacks were wrongly classified as IPsweep attacks. This could be attributed to the fact that the bootstrap samples that were generated randomly may contain instances that might be biased towards IPsweep attacks. The bar graph for all attack accuracy is depicted in Fig. 10.22.

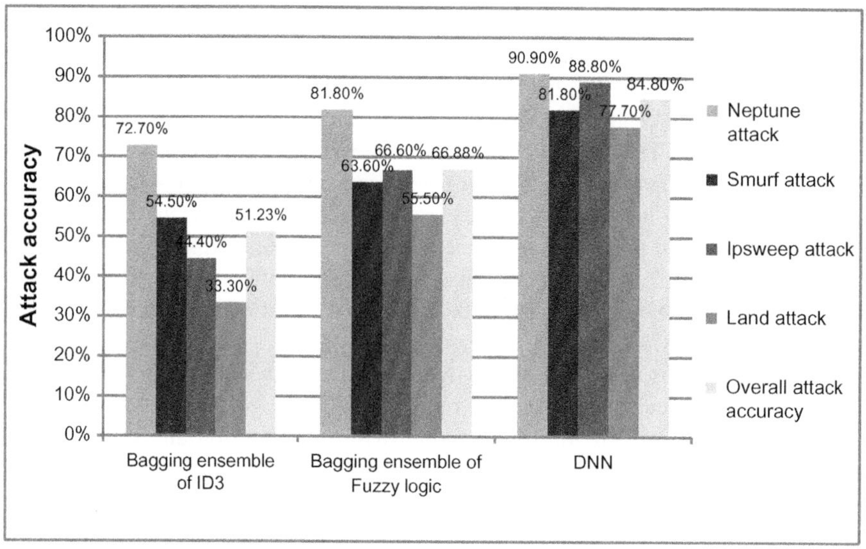

FIG. 10.22

Accuracy rate by algorithms—Bagging ensemble of ID3, bagging ensemble of fuzzy logic, and DNN.

Among the 50 packets in the test set, 11 packets were Neptune packets. Bagging Ensemble of ID3 and Bagging Ensemble of Fuzzy Logic detected eight and nine packets as Neptune among the 11 Neptune packets, whereas DNN detected 10 packets. As shown inTable 4.7, the Neptune attack accuracy for DNN is 0.909, which is higher than the Neptune attack accuracy for Bagging Ensemble of ID3 having a value of 0.727 and Bagging Ensemble of Fuzzy Logic having a value of 0.818. The bar graph for Neptune attack accuracy is depicted in Fig. 10.22.

Among the 50 packets in the test set, nine packets were IPsweep packets. Bagging Ensemble of ID3 and Bagging Ensemble of Fuzzy Logic detected four and six packets as IPsweep among the 11 IPsweep packets, whereas DNN detected eight packets as IPsweep. Hence the IPsweep attack accuracy for DNN is 0.888, which is higher than the IPsweep attack accuracy for Bagging Ensemble of ID3 having a value of 0.444 and Bagging Ensemble of Fuzzy Logic having a value of 0.888.

Among the 50 packets in the test set, nine packets were Land packets. Bagging Ensemble of ID3 and Bagging Ensemble of Fuzzy Logic detected three and five packets as Land among the 11 Land packets, whereas DNN detected seven packets as Land. The Land attack accuracy for DNN is 0.777, which is higher than the Land attack accuracy for Bagging Ensemble of ID3 having a value of 0.333 and Bagging Ensemble of Fuzzy Logic having a value of 0.565. The bar graph for Land attack accuracy is depicted in Fig. 10.22.

Among the attacks sent, DNN's attack accuracy is higher compared to Bagging Ensemble of ID3 and Bagging Ensemble of Fuzzy Logic. Bagging Ensemble of ID3's attack accuracy is low for all types of attacks that were sent. Bagging Ensemble of Fuzzy Logic detects attacks better than Bagging Ensemble of ID3. Among these attacks, DNN can easily detect Neptune and IPsweep attacks compared to Smurf and Land attacks. Both Bagging Ensemble of ID3 and Bagging Ensemble of Fuzzy Logic can easily detect Neptune attacks compared to all other attacks.

Fig. 10.21 depicts a bar graph of overall accuracy for each of the three algorithms. The overall accuracy for DNN is 92%, which is higher than Bagging Ensemble of ID3 with overall accurace of 70% and Bagging Ensemble of Fuzzy Logic with overall accuracy of 82%. FAR for DNN is 0.026, which is less Bagging Ensemble of ID3 with a FAR of 0.12 and Bagging Ensemble of Fuzzy with a FAR of 0.102. ADR for DNN is 0.925, which is higher than Bagging Ensemble of ID3 with an ADR of 0.725 and Bagging Ensemble of Fuzzy Logic with an ADR of 0.875. Finally, attack accuracy of DNN is 0.791, which is higher than Bagging Ensemble of ID3 with an attack accuracy of 0.583 and Bagging Ensemble of Fuzzy Logic with an attack accuracy of 0.66. In all cases, DNN performs better than the other two algorithms.

In this work, we proposed a better approach for replacing rules in the flow table that reduced the number of flow modification messages sent to the switches. Several test sets were used in performance evaluation of our Optimal Rule Replacement Manager. The results are interesting in terms of the good behavior of the algorithms. This is due to replacement of rules even before the flow table entry becomes full. Security challenges related to malicious attacks were dealt. For this, we implemented an IDS that uses machine-learning techniques to detect security-related attacks in the network. This IDS was implemented as a module in the controller.

10.5 Conclusion

SDN is a networking paradigm that is moving at a fast pace. It simplifies network configuration and management and makes the network evolvable. Decoupling the data plane from the control plane offers high flexibility. The code developed and tested on Mininet, for an OpenFlow controller, modified switch, or host, can be easily moved to a real system with minimal changes for real-world testing, performance evaluation, and deployment. In this work, apart from flow control and flow management in a home network using SDN, an IDS was implemented at the controller to detect attacks by an intruder. The advantage of this approach is that any new packets that are sent to the SDN-enabled devices in multihomed networks must go through the controller any IoT-enabled SDN setup. The IDS implemented in the controller can detect if a packet is malicious. If so, the controller discards the packet thus preventing an attack.

Bagging Ensemble of ID3, Bagging Ensemble of Fuzzy Logic, and DNN algorithms were used in implementing the proposed IDS. A test set consisting of 50 packets were sent in real time for evaluating the performance of the algorithms. Out of these 50 packets, 40 were malicious and 10 were normal. FAR for DNN is 0.026, which is less than Bagging Ensemble of ID3with a FAR of 0.121 and Bagging Ensemble of Fuzzy Logic with a FAR of 0.102. ADR for DNN is 0.925, which is higher Bagging Ensemble of ID3 with an ADR of 0.725 and Bagging Ensemble of Fuzzy Logic with an ADR of 0.875. Attack accuracy of DNN is 0.791, which is higher than Bagging Ensemble of ID3 with attack accuracy of 0.583 and Bagging Ensemble of Fuzzy Logic with an attack accuracy of 0.66. The overall accuracy for DNN is 92%, which is higher than Bagging Ensemble of ID3 with overall accuracy of 70% and Bagging Ensemble of Fuzzy Logic with overall accuracy of 82%. In all cases, DNN performs better than the other two algorithms.

Appendix 1: Implementation screenshots

Network is created using Miniedit GUI. The sample network we created consisted of 16 hosts, two switches, and one controller. This is depicted in Fig. A1.1. All hosts can be manually configured. Configuration includes IP address, link bandwidth, jitter, latency, and so on.

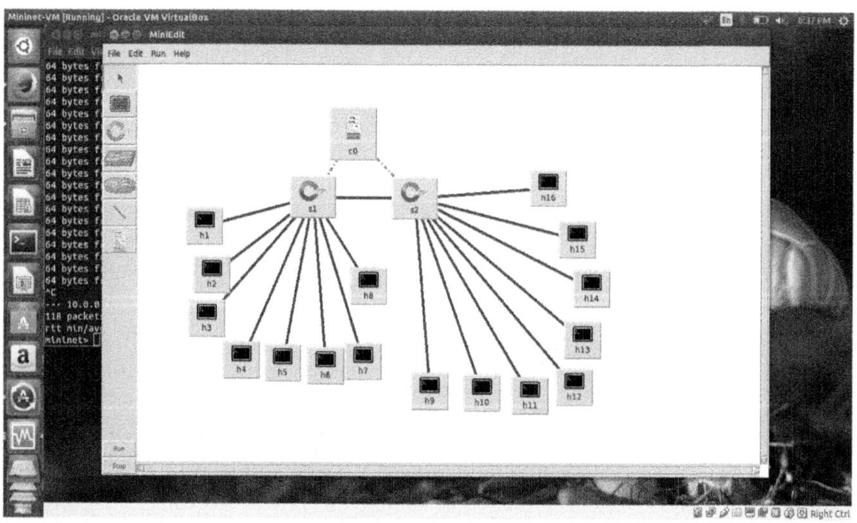

FIG. A1.1

Creation of network.

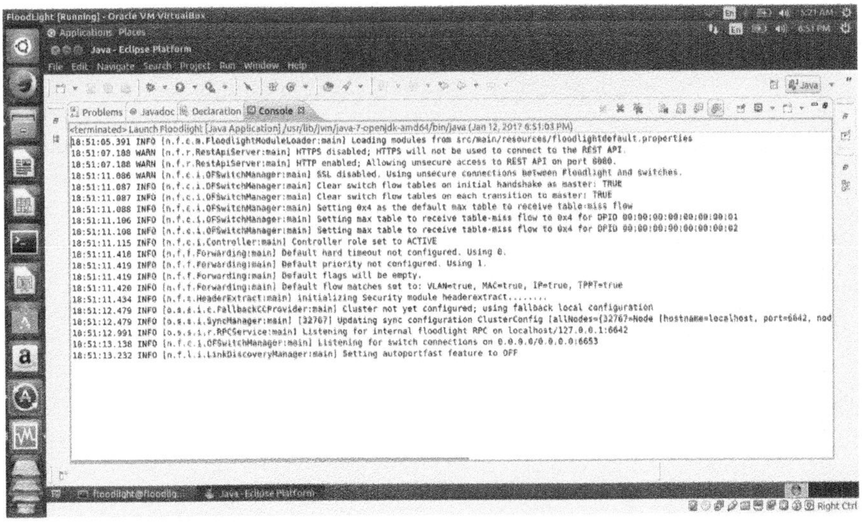

FIG. A1.2

Connection establishment and security module initialization.

The successful connection establishment and security initialized with initial configuration parameters are depicted in Fig. A1.2. Modules are initialized in the order in which they occur in the Floodlight default properties file.

The packets are sent between hosts using a ping command. You can also specify the number of packets to be sent between hosts. The "pingall" command is used to check host reachability from one host to every other host in the network. Fig. A1.3 shows the output of the ping and pingall commands.

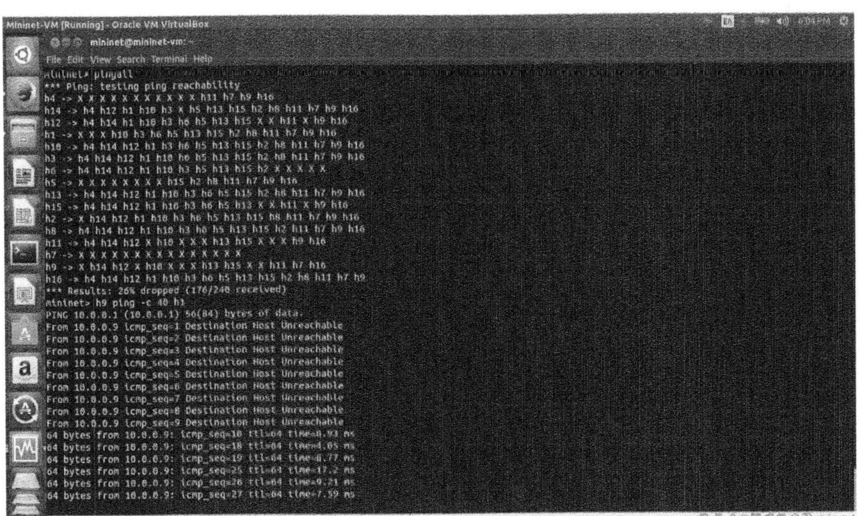

FIG. A1.3

Sending packets between hosts.

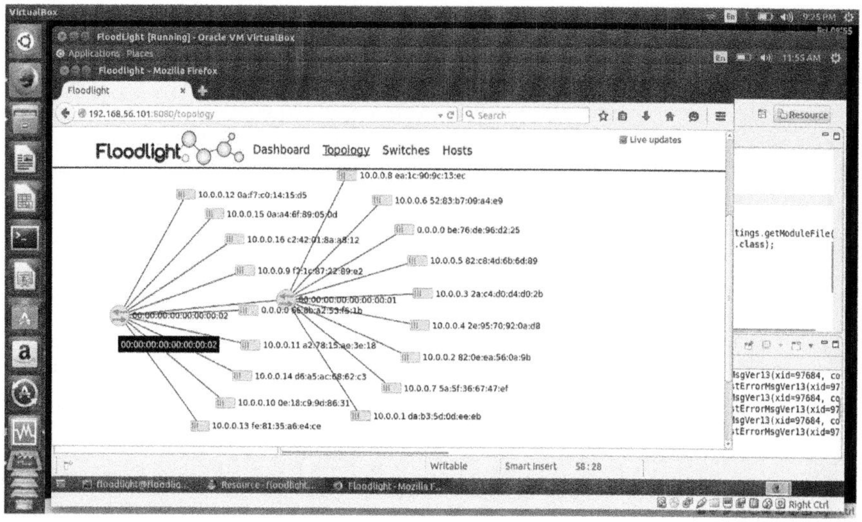

FIG. A1.4

Dynamic network creation at controller side.

The Floodlight controller creates a network of hosts as when it receives requests from different hosts dynamically. This is a web-based GUI that can display the network of hosts and switches, host details and switch flow details. This is depicted in Fig. A1.4.

The header information of the packets that is sent to the controller can be extracted by implementing a header extract sub module in the security module. Fig. A1.5 depicts the successful header information extraction at the controller side

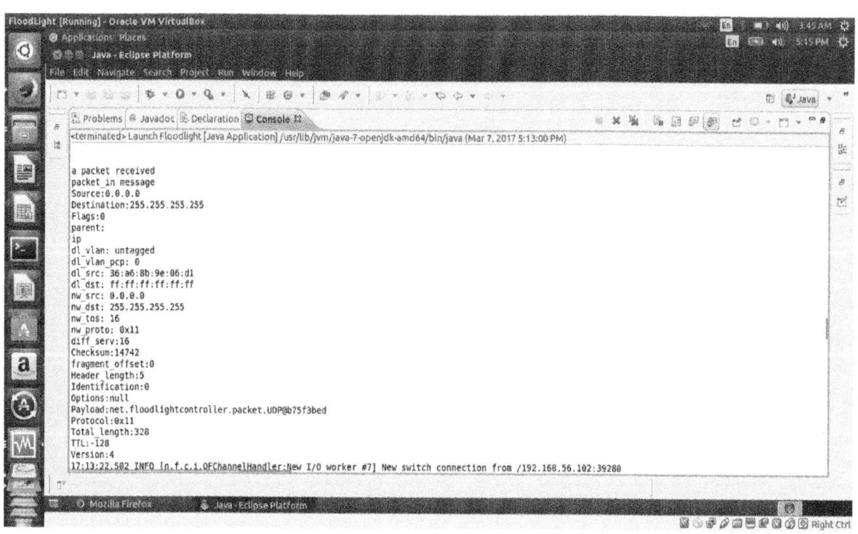

FIG. A1.5

Packet extraction in the security module at the controller side.

by the security module. The features that were extracted include SRC_IP, DEST_IP, IP Flags, TOP, TOS, fragment offset, IP Header Length, total IP packet length, TTL, and version. If the packet is a TCP packet, TCP flags and urgent field are extracted. Among those, only certain features were sent to the IDS module for detection. SRC_IP and DEST_IP were used to calculate land field. IP header length and total IP packet length were used to calculate the source bytes.

In the learning phase, bootstrap samples had to be generated for Bagging Ensemble of ID3 classifiers. Fig. A1.6 depicts the successful generation of bootstrap samples for Bagging_ID3 sub module. Three bootstrap samples were generated and used for creating models. A similar process takes place at the Bagging_FuzzyLogic sub module.

The predictions given by each model will have to be aggregated and maximal voting has to be done to find the final prediction for a given packet. Fig. A1.7 depicts the bootstrap aggregation for Bagging_ID3 sub module. The predictions given by each model and the maximal vote are also displayed. A similar process takes place at the Bagging_FuzzyLogic sub module.

Rules have to be generated for the models in the Bagging_FuzzyLogic sub module. Fig. A1.8 depicts the successful generation of rules for model 1 by the FURIA algorithm. Similar rules are generated for the other models based on the bootstrap samples of each model.

The DNN has to be trained using the given dataset. Fig. A1.9 depicts the learning rate and cost during each iteration in the DNN sub module. The learning rate value was fixed to 1. The number of hidden layers of the convolution network was set to three. The iteration continues until the cost converges.

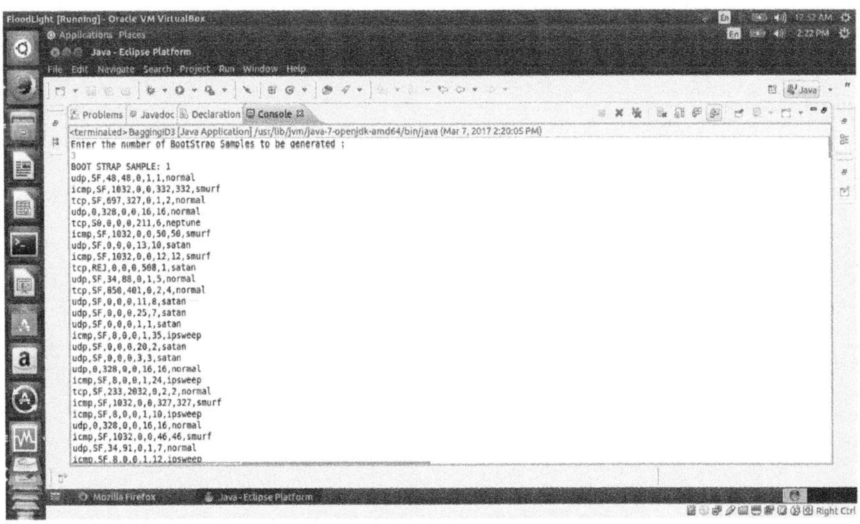

FIG. A1.6

Bootstrap sample generation in Bagging_ID3 sub module.

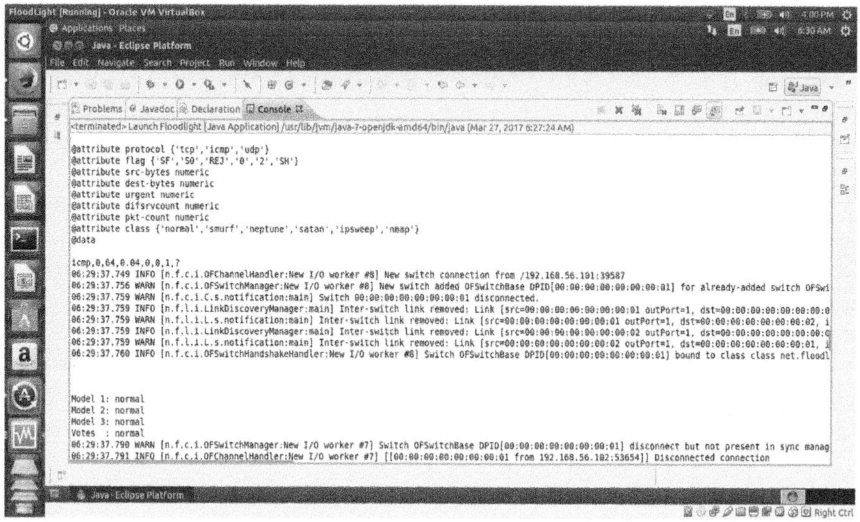

FIG. A1.7

Bootstrap aggregation for Bagging_ID3 sub module.

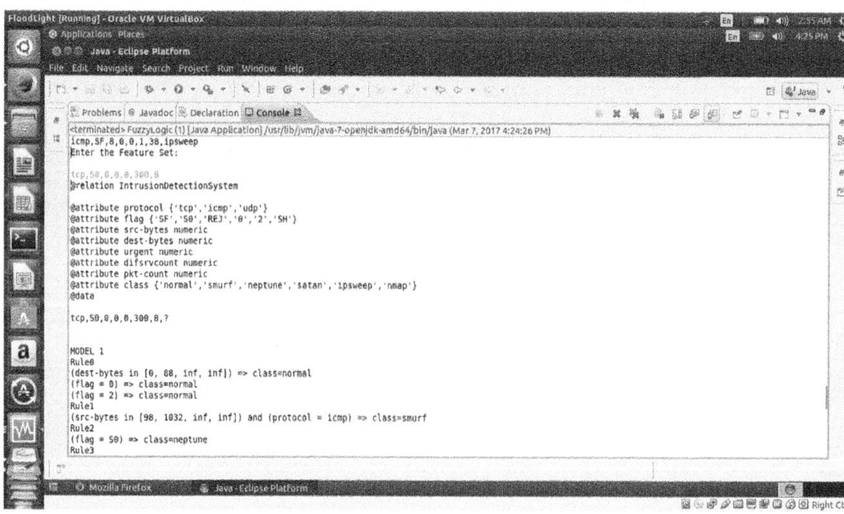

FIG. A1.8

Fuzzy rule creation in Bagging_FuzzyLogic sub module.

A Smurf attack can be sent by flooding ICMP packets to the victim IP. Fig. A1.10 illustrates how a Smurf attack is sent from Mininet using the hping3 tool. Since it is DDoS attack, the broadcast address must be specified. "−1" denotes that it is an

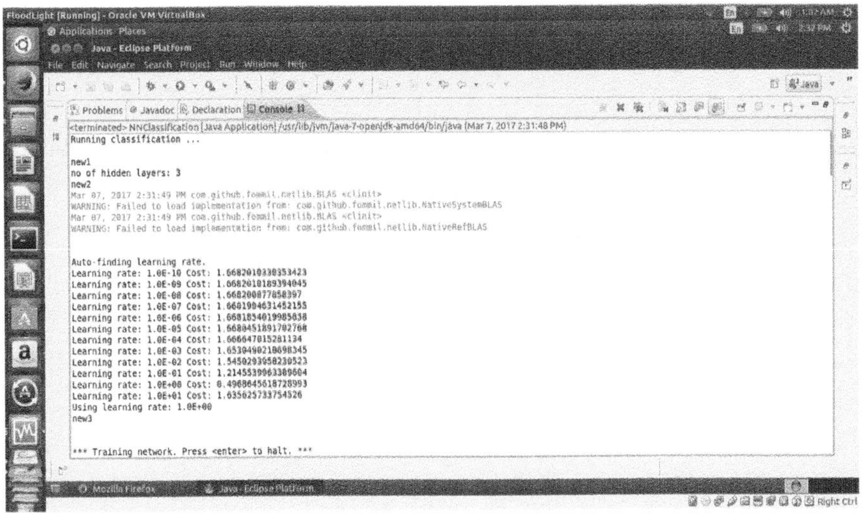

FIG. A1.9

Training neural network in DNN sub module.

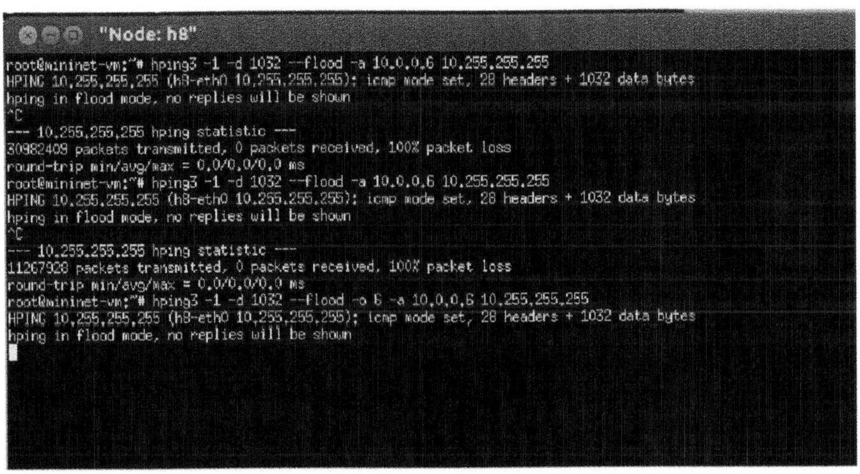

FIG. A1.10

Sending a Smurf attack from Mininet using hping3.

ICMP packet. "Flood" should be specified to flood the packets. Victim IP must be specified to send an attack. Additionally data packet size and type of service field can also be changed. For a Smurf attack, the data packet size is large.

Fig. A1.11 shows the header information when a Smurf attack was performed. The data shows that it is an ICMP packet, the victim IP is 10.0.0.6, broadcast address is 10.255.255.255, and data packet size is 1040 bytes. This attack sends an ICMP

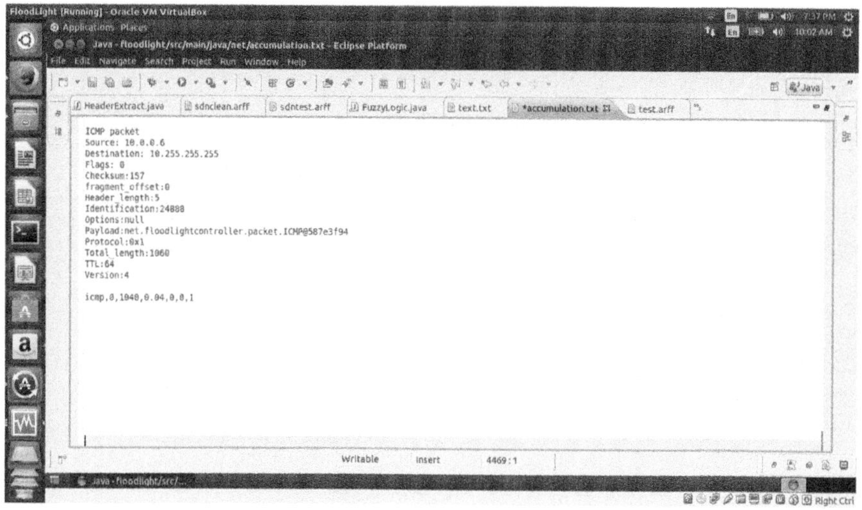

FIG. A1.11

Header details extracted for a Smurf attack.

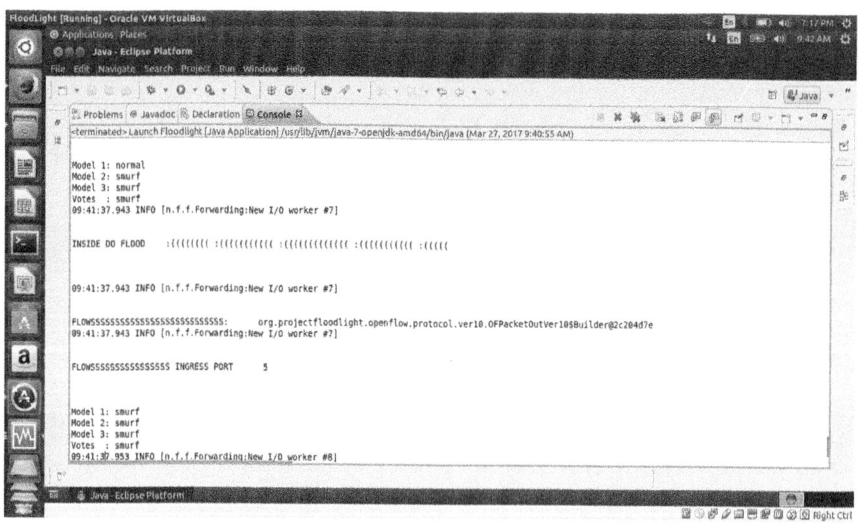

FIG. A1.12

Smurf attack detection by the security module at the controller side.

echo request to the specified broadcast address. The active hosts in the broadcast address sends ICMP echo reply to the victim IP thus performing a DDos attack.

Fig. A1.12 depicts the detection of a Smurf attack by the Bagging_FuzzyLogic sub module. Bootstrap aggregation and maximal voting takes place in deciding the final prediction.

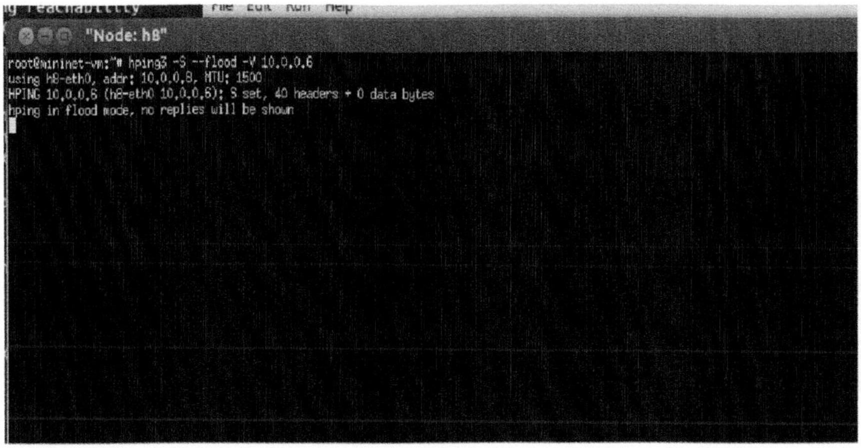

FIG. A1.13

Sending a Neptune attack from Mininet using hping3.

A Neptune attack is sent by flooding TCP SYN packets to the victim's IP. Fig. A1.13 illustrates how a Neptune attack is sent using the hping3 tool. The SYN field is set to 1 by specifying "–S" in the command line. "Flood" denotes that packets are flooded. The victim IP is 10.0.0.6. This command floods TCP SYN packets to the same IP (10.0.0.6) thus performing a DoS attack.

Fig. A1.14 depicts the detection of a Neptune attack by the Bagging_FuzzyLogic sub module. Bootstrap aggregation and maximal voting takes place in deciding the

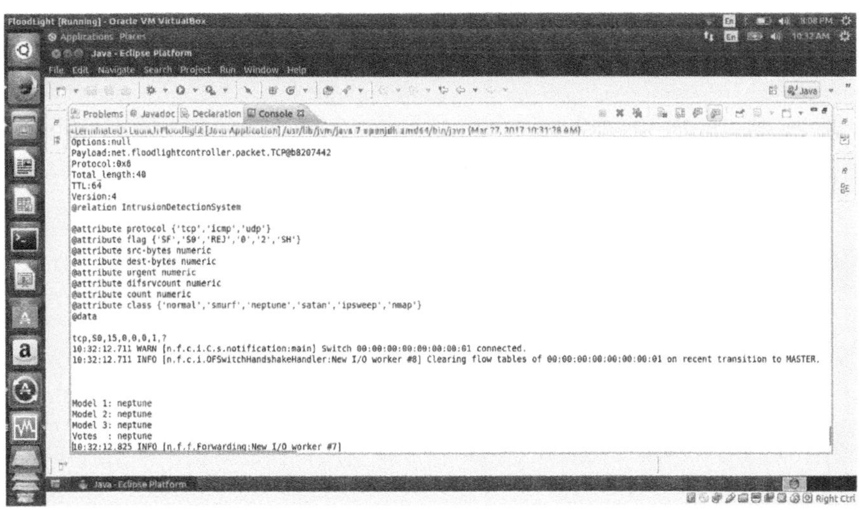

FIG. A1.14

Neptune attack detection by the security module at the controller side.

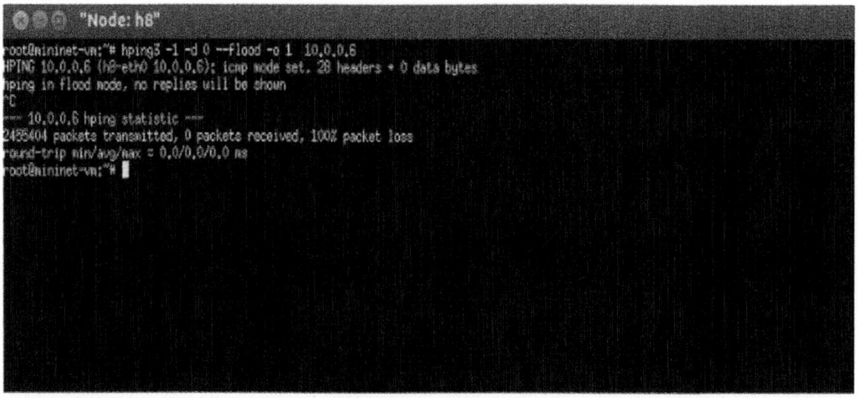

FIG. A1.15

IPsweep attack sent from Mininet using hping3.

final prediction. From the screenshot it can noted that the packet is TCP and the flag is S0, indicating that it is a SYN flood/Neptune attack.

An IPsweep attack is sent using an ICMP echo request packet with zero data size. Fig. A1.15 depicts how an IPsweep attack is sent from Mininet using the hping3 tool. "−1" indicates that it is an ICMP packet. Data size if set to 0 for an IPsweep attack. The differential service is set to 1, which means the packet should be immediately delivered.

Fig. A1.16 depicts the detection of an IPsweep attack by the Bagging_FuzzyLogic sub module at the controller side. The packet extracted is an ICMP packet with

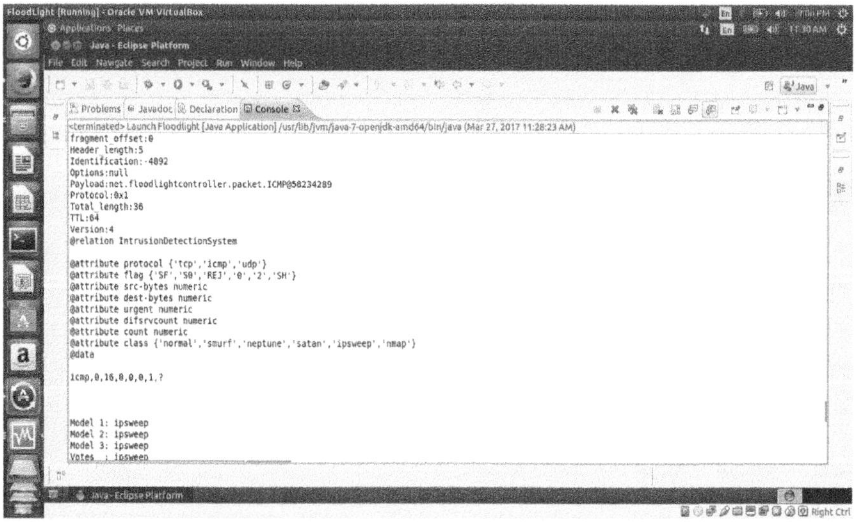

FIG. A1.16

IPsweep attack detection by the security module at the controller side.

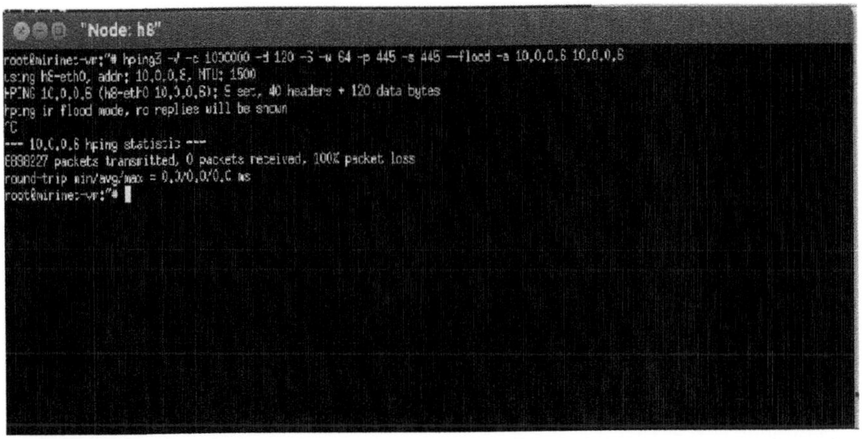

FIG. A1.17

Land attack sent from Mininet using hping3.

zero source bytes and a diffserv value of 1. Bootstrap aggregation and maximal voting takes place to decide on a final prediction for a packet.

A Land attack is a spoofed DoS attack where the SRC_IP and DEST_IP are the same. Fig. A1.17 depicts how a Land attack is sent from Mininet using the hping3 tool. The spoofed SRC_IP is set using "–a" in the command line. "–d" is used to specify the data size. "Flood" is used to flood the packets. Victim IP is 10.0.0.6. The SRC_IP is spoofed to 10.0.0.6, mimicking a Land attack.

Fig. A1.18 depicts the detection of a Land attack by the security module at the controller side. The SRC_IP and DEST_IP have the same address of 10.0.0.6.

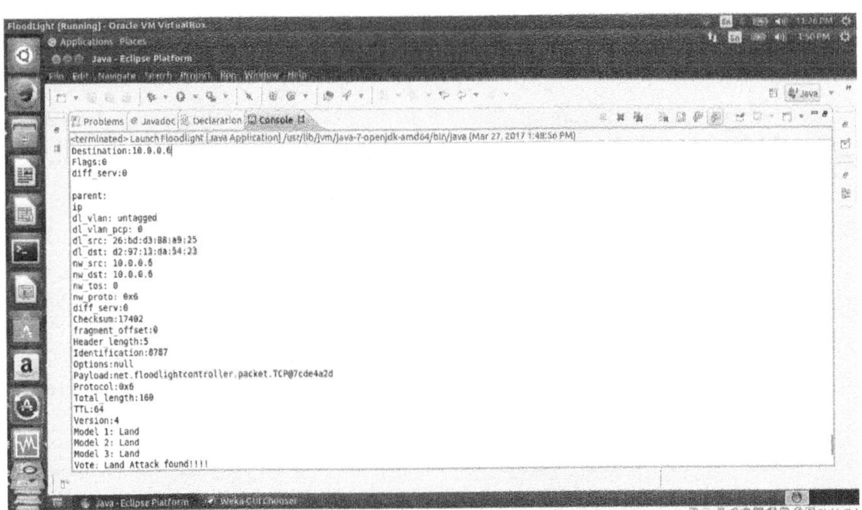

FIG. A1.18

Land attack detection by the security module at the controller side.

References

[1] J. Jo, S. Lee, J.W. Kim, Software-defined home networking devices for multi-home visual sharing, IEEE Trans. Consum. Electron. 60 (3) (2014) 534–539.

[2] R. Mortier, T. Rodden, T. Lodge, D. McAuley, C. Rotsos, A.W. Moore, A. Koliousis, J. Sventek, Control and understanding: owning your home network, in: 2012 Fourth International Conference on Communication Systems and Networks (COMSNETS), January, IEEE, 2012, pp. 1–10.

[3] S. Yoon, S.S. Lee, S.H. Kim, Seamless and secure P2P communication scheme for mobile Internet devices behind NAT, in: 2012 IEEE International Conference on Consumer Electronics (ICCE), January, IEEE, 2012, pp. 347–348.

[4] M. Boussard, D.T. Bui, L. Ciavaglia, R. Douville, M. Le Pallec, N. Le Sauze, L. Noirie, S. Papillon, P. Peloso, F. Santoro, Software-defined LANs for interconnected smart environment, in: 2015 27th International Teletraffic Congress (ITC 27), September, IEEE, 2015, pp. 219–227.

[5] S. Notra, M. Siddiqi, H.H. Gharakheili, V. Sivaraman, R. Boreli, An experimental study of security and privacy risks with emerging household appliances, in: 2014 IEEE Conference on Communications and Network Security (CNS), October, IEEE, 2014, pp. 79–84.

[6] J. Joshi, V. Rajapriya, S.R. Rahul, P. Kumar, S. Polepally, R. Samineni, D.K. Tej, Performance enhancement and IoT based monitoring for smart home, in: 2017 International Conference on Information Networking (ICOIN), January, IEEE, 2017, pp. 468–473.

[7] L. Ying-Tsung, H. Wei-Hsuan, H. Chin-Meng, T.C. Seng-Cho, An integrated cloud-based smart home management system with community hierarchy, IEEE Trans. Consumer Device 62 (2016) 1–9.

[8] N. Feamster, Outsourcing home network security, in: ACM SIGCOMM Workshop on Home Networks, 2010, pp. 37–42.

[9] K. Tom, A. Django, D. Karim, J. Ming, Risk driven Smart Home resource management using cloud services, in: Future Generation Computer Systems, 30, Elsevier, 2014, pp. 13–22.

[10] C. Pradhan, C. Pradhan, H. Das, B. Naik, N. Dey, Handbook of Research on Information Security in Biomedical Signal Processing, (2018).

[11] B.S.P. Mishra, H. Das, S. Dehuri, A.K. Jagadev (Eds.), Cloud Computing for Optimization: Foundations, Applications, and Challenges, Springer International Publishing, 2018.

[12] B.S. Lee, R. Kanagavelu, K.M.M. Aung, An efficient flow cache algorithm with improved fairness in software-defined data center networks, in: 2013 IEEE 2nd International Conference on Cloud Networking (CloudNet), November, IEEE, 2013, pp. 18–24.

[13] D. Kreutz, E.P. Verı´ssimo, S. Azodolmolky, Software-defined networking: a comprehensive survey, in: Proc. IEEE International Conference on SDN, 103, 2015, pp. 10–13.

[14] S. Elhag, F. Alberto, B. Abdullah, A. Saleh, H. Francisco, On the combination of genetic fuzzy systems and pairwise learning for improving detection rates on intrusion detection systems, Int. J. Expert Syst. Appl. 42 (2015) 193–202.

[15] R. Devi, R.K. Jha, A. Gupta, S. Jain, P. Kumar, Implementation of intrusion detection system using adaptive neuro-fuzzy inference system for 5G wireless communication network, AEU Int. J. Electron. Commun. 74 (2017) 94–106.

[16] D.P. Gaikwad, C.T. Ravindra, Intrusion detection system using bagging ensemble method of machine learning, in: Proc. IEEE Conference on Computing Communication Control and Automation, 2014, pp. 291–295.

[17] A.T. Tuan, M. Lotfi, M. Des, A.R. Syed, G. Mounir, Deep learning approach for network intrusion detection in software defined networking, in: Proc. IEEE International Conference on Wireless Networks and Mobile Communications (WINCOM), 2016, pp. 258–263.

[18] M. Zhu, K. Ye, C.Z. Xu, Network anomaly detection and identification based on deep learning methods, in: International Conference on Cloud Computing, June, Springer, Cham, 2018, pp. 219–234.

[19] N. Shone, T.N. Ngoc, V.D. Phai, Q. Shi, A deep learning approach to network intrusion detection, IEEE Trans. Emerg. Top. Comput. Intell. 2 (1) (2018) 41–50.

[20] J. Li, Z. Zhao, R. Li, H. Zhang, T. Zhang, AI-based two-stage intrusion detection for software defined IoT networks, IEEE Internet Things J. 6 (2) (2018) 2093–2102.

[21] A.C. Jonathan, D.J. Conrad, B.A. James, A signal processing approach for cyber data classification with deep neural networks, Elsevier J. Cyber Secur. 61 (2015) 349–354.

[22] C.R. Panigrahi, M. Tiwary, B. Pati, H. Das, Big data and cyber foraging: future scope and challenges, in: Techniques and Environments for Big Data Analysis, Springer, Cham, 2016, pp. 75–100.

[23] K. Slavov, M. Pourzandi, D. Migault, Identifying and addressing the vulnerabilities and security issues of SDN, Ericson Technol. Rev. 62 (7) (2015) 1–11.

[24] J.A. Cox, C.D. James, J.B. Aimone, A signal processing approach for cyber data classification with deep neural networks, Procedia Comput. Sci. 61 (2015) 349–354.

[25] M.J. Kang, j.W. Kang, Novel intrusion detection methos using deep neural network for in-vehicle network security, in: Proc. IEEE International Conference on Vehicular Technology, 2016, pp. 1–5.

[26] D. Kreutz, F.M.V. Ramos, P.E. Veríssimo, C.E. Rothenberg, S. Azodolmolky, S. Uhlig, Software-defined networking: a comprehensive survey, Proc. IEEE 103 (2015) 10–13.

[27] N. Masahide, F. Yusuke, I. Hiroshi, M. Kenichi, Implementing multi-vendor home network system with vendor neutral services and dynamic service binding, in: Proc. IEEE International Conference on Services Computing, 2, 2008, pp. 275–282.

[28] D. Sergei, V. Andrei, L. Ivan, A fuzzy logic-based information security management for software-defined networks, in: Proc. IEEE International Conference on Advanced Communication Technology, 2014, pp. 167–171.

[29] H. Taein, P. Hojin, W.C. Jin, Personal mobile A/V control point for home-to-home media streaming, in: Proc. IEEE International Conference on Consumer Electronics, 2008, pp. 1–2.

[30] S. Mourougan, M. Aramudhan, Hybrid evolutionary algorithmn based intrusion detection system for denial of service attacks. Indian J. Sci. Technol. 8 (35) (2015), https://doi.org/10.17485/ijst/2015/v8i35/86652.

[31] Octopress, www.docs.mininet.org, 2015.

[32] Project Floodlight, http://www.projectfloodlight.org/floodlight, 2010.

[33] Salvatore Sanfilippo, http://www.hping.org/, 2006.

[34] M. Latah, L. Toker, Towards an Efficient Anomaly-Based Intrusion Detection for Software-Defined Networks, 2018. arXiv preprint arXiv:1803.06762.

[35] P. Mishra, V. Varadharajan, U. Tupakula, E.S. Pilli, A detailed investigation and analysis of using machine learning techniques for intrusion detection, IEEE Commun. Survey Tut. 21 (1) (2018) 686–728.

[36] I. Chopra, M. Singh, SHAPE—an approach for self-healing and self-protection in complex distributed networks, J. Supercomput. 67 (2) (2014) 585–613.
[37] P.T. Jagannath, To Improve the Performance of Deep Packet Inspection for Network Intrusion Detection, Guru Gobind Singh Indraprastha University, 2015.
[38] J. Cui, J. He, Y. Xu, H. Zhong, TDDAD: time-based detection and defense scheme against DDoS attack on SDN controller, in: Australasian Conference on Information Security and Privacy, July, Springer, Cham, 2018, pp. 649–665.
[39] P. Wang, K.M. Chao, H.C. Lin, W.H. Lin, C.C. Lo, An efficient flow control approach for SDN-based network threat detection and migration using support vector machine, in: 2016 IEEE 13th International Conference on e-Business Engineering (ICEBE), November, IEEE, 2016, pp. 56–63.
[40] D. Bolzoni, S. Etalle, Approaches in anomaly-based network intrusion detection systems, Intrus. Detect. Syst. 38 (2008) 1–15.
[41] A. Balk, M. Gerla, D. Maggiorini, M. Sanadidi, Adaptive video streaming: pre-encoded MPEG-4 with bandwidth scaling, Comput. Netw. 44 (4) (2004) 415–439.

Index

Note: Page numbers followed by *f* indicate figures, *t* indicate tables, and *b* indicate boxes.

D

G

H

I

K

L

R

CPI Antony Rowe
Eastbourne, UK
November 20, 2019